Physics for Nonphysicists

FRANK R. SPELLMAN

Government Institutes
An imprint of
The Scarecrow Press, Inc.
Lanham, Maryland • Toronto • Oxford
2009

Published in the United States of America
by Government Institutes, an imprint of The Scarecrow Press, Inc.
A wholly owned subsidiary of
The Rowman & Littlefield Publishing Group, Inc.
4501 Forbes Boulevard, Suite 200
Lanham, Maryland 20706
http://govinst.govinstpress.com

Estover Road
Plymouth PL6 7PY
United Kingdom

British Library Cataloguing in Publication Information Available

Library of Congress Cataloging-in-Publication Data
Spellman, Frank R.
  Physics for nonphysicists / Frank R. Spellman.
    p.   cm.
  Includes bibliographical references and index.
  ISBN 978-0-86587-183-0 (pbk. : alk. paper) — ISBN 978-1-60590-279-1
(electronic)
  1. Physics—Popular works.   I. Title.
  QC24.5.S64   2009
  530—dc22   2008050372

∞ ™ The paper used in this publication meets the minimum requirements
of American National Standard for Information Sciences—Permanence of
Paper for Printed Library Materials, ANSI/NISO Z39.48-1992.
Manufactured in the United States of America.

For
Kimberly Ann Barnes

Contents

Preface

Why a text on physics for nonphysicists? Answer: Based on personal experience, I have found that many professionals do have some background in physics (don't we all?), but many of these folks need to stay current; this is one of the purposes of this text. Moreover, many environmental practitioners are specialists (e.g., epidemiologists, water quality technicians, community health specialists, toxicologists, environmental science and occupational health and safety professionals, industrial hygienists, teachers, and so forth). Herein lies the problem—that is, specialization. My view is that practitioners of any type should know their field of specialization but also should be generalists with a wide range of knowledge—not just specialists whose knowledge may be too narrowly focused. Again, based on personal experience, my students who have generalized their education (spread out their exposure to include disciplines in several aspects of environmental studies), versus those who narrowly specialize, are afforded many more opportunities to broaden their upward mobility in employment ventures; they have a much better chance to rise to upper management positions, where more prestige, responsibility, pecuniary rewards, and, yes, more headaches await them. It is true that every problem has a solution, but determining the solution is not always easy. A proper, wide-ranging education makes problem-solving easier, however.

Physics for nonphysicists? Isn't that an oxymoron, like "same difference" or "bicycle safety" or "sanitary sewer"? Or is this title used here to add humor like Richard Feynman's use (in his highly regarded and renowned lectures on physics) of the deliberate oxymoron "dry water," adding humor to his otherwise-dry analysis of the hydrodynamics of fluids? Not really. Although physics certainly can be one of those narrow disciplines mentioned above and some consider the topic "dry," it is one of those disciplines that explains, in simple (general) terms, how the physical world works—and where it applies to all of the nonphysicists in this world (and our number is too large to count).

Physics for Nonphysicists provides an overview and review of the important aspects of physics in a way that can be easily understood—even if the user has not taken any formal physics courses.

Along with basic physics principles, the text provides a clear, concise presentation of the consequences of physics' interaction with the environment we inhabit. Although *Physics for Nonphysicists* is primarily designed for professionals (especially practicing environmental professionals) who need a quick review of physics principles and applications, it is also designed for a more general audience. Even if you are not tied to a desk, I provide you, the nonphysicist, with the jargon, concepts, and key concerns of physics and physics in action. Starting with the very basic principles of measurement, conversion factors, and math operations, we will explore the topics of motion and force, work and energy, gravity, atoms, heat, sound, and light and color, ending with a detailed discussion of basic electricity. This book is compiled in an accessible, user-friendly format, unique in that it explains scientific concepts in the most basic way possible.

This book is designed to serve the needs of students, teachers, consultants, and technical personnel who need to review the principles, basics, and fundamentals of physics. The text is packed with information students need to get ready for SAT exams. In order to maximize the usefulness of the material contained in the text, I have written and presented it in plain English and in a simplified and concise format.

Each chapter ends with a chapter review test to help evaluate mastery of the concepts presented. Before going on to the next chapter, you should take the review test, compare your answers to the key provided in the appendix, and review the pertinent information for any problems you missed. If you miss many items, review the whole chapter.

Again, this text is accessible to those who have no experience with physics. If you work through the text systematically, you will acquire an understanding of and skill in the principles of physics—adding a critical component to your professional knowledge.

Frank R. Spellman
Norfolk, Virginia

Introduction

Why study physics? This is a common question from many of my undergraduate students majoring in environmental health and safety, occupational health, hazardous materials management, risk management, and/or industrial hygiene. I hear this question often, especially at the beginning of their studies. Generally, they seem to understand that the study of human anatomy and physiology, chemistry, biology, math, and other elemental topics is important to them because they relate to their major areas of study. But physics? No way!

When I ask students why they think the study of physics is not important to them or related to the disciplines they are majoring in, they give different answers. The most common answer, however, is more related to the hard work required to successfully complete college physics and receive a grade of B or higher. Common statements that I have heard about having to take physics include: too much math; too much homework; too much science; too much work; a real pain in the %$#(@; et cetera, et cetera, et cetera. "Physics? No way!"

One school day, at the beginning of one of my environmental health classes, I was rudely awakened (startled) back from my academic la-la-land to the real world around me (according to my students, classrooms are generally touted as real-world locations but under the tutelage of out-of-this-world weirdos; they informed me that these words of wisdom did not necessarily apply to me) by a heavy, loud bang. One of my young male students had taken his five-pound, 1,503-page physics textbook, raised it high above his head, and dropped it on his desk. (Immediately Ted Nugent's refrain ran through my ancient brain cells: "Man, if it's too loud, you're too old.") Anyway, the student, who now had the full attention of a class of forty-three other students and one ancient non-weirdo academic, stated, "Man, not only is this book at $180 too expensive—man, it's heavy as hell! Physics? No way, man!" Just about all of the other forty-three students either nodded their heads in agreement, yelled a profound "Amen!" or both.

Of course, after the book-hitting-the-desk rude awakening and the couched comments about physics being heavy, it was not long (seconds actually) before the students—all forty-four of them—were sitting in their desks, quiet and all looking directly at me. They were waiting for me to counter the young man's comments about the heaviness (and therefore the hinted needlessness) of physics in general. Most of these students knew me (they had had classes with me before) and knew that I had comments about anything and everything and that I was not afraid to speak up and out—not even afraid to counterattack such disparaging comments about such a lofty science. So I did; I did not disappoint them. No, not at all.

Glancing around the room, I began my normal selling approach; that is, initially

I agreed with them (it is important to make them think you are on their side of any argument). I simply told them that learning physics is hard work; it is a science that requires concentration and study. And when I used to be a student, sitting out there in the classroom like they were/are, I asked myself the same question: "Why physics?"

Then I went on to say that after graduating from college and working in the real world for several years as an environmental safety and health professional, I quickly learned the value of physics—the value of knowing its rules and principles and applications. I quickly learned that physics is crucial to gaining an understanding of not only the world around us but also the world inside us (the anatomy and physiology of the human body). I explained that physics is the basis of many other sciences and scientific phenomena, including astronomy, oceanography, chemistry, seismology, ergonomics, mechanics, and mechanical advantage. Moreover, through practice I found that physics is not limited to the "hard sciences." Increasingly, physicists are turning their expertise to biochemistry and biology itself. I told them that over the years as a professor I had had pre-med students who came back to me later and told me how grateful they were that they had learned the basics of physics at this level of school—and yes, physics is important even in medicine.

After presenting general, non-window-dressing-type information, I then laid on the heavy stuff—the attention-getters. I explained that without physics they would have no automobiles to use to get to and from class and elsewhere. Without physics, there would be no cell phones—no text messaging. Without physics they would have no television, no video games, and so forth. At this point I had their full, undivided attention. I went on: Physics prepares us all for what is ahead of us in our increasingly technical world.

Then I finished with my bottom line (I always feel it is important to deflate the balloon of discontent at the end of a diatribe): "Remember, dear students of mine, in this program, physics is required to get your bachelor of science degree."

The Purpose of This Text

As stated in the preface, this book is not intended to make physicists out of those who read it. Only through extensive classroom training and practical field experience can you accomplish this. Instead, *Physics for Nonphysicists* can help you become familiar with the world around you. Based on the constructive criticism of the first three successful volumes of this series, *Chemistry for Nonchemists*, *Biology for Nonbiologists*, and *Ecology for Non-Ecologists*, many readers said that personal curiosity and a desire to learn new things drove them to use and study these texts.

In this book, we begin our exploration of physics with very basic topics and progress to topics that are more complex—fully explained so as to be easily understood. We start with an explanation of measurement, conversion factors, and basic math concepts, and end with a comprehensive overview of electricity.

In perusing this text it is important to keep the following in mind: Physics is not a dead science; rather, it is a living science—literally a work in progress. This is clearly demonstrated on a daily basis by the introduction of new and exciting technological discoveries and innovations. Many new laws and principles of physics begin as hypotheses (as is required by the tenets of the scientific method) and eventually lead us to proven theories that have real-world practical application, making our lives on Earth more comfortable and also more interesting. Thus in regard to the study of physics, many of us might state: "Physics? Physics is OK, man!"

CHAPTER 1

Measurement

Like much of science (and life itself), physics is all about measurement.

Units of Measurement

We do not go through a day without measuring something. Whether counting the money in our pockets or in our bank accounts or in our 401(k)s, counting the number of shopping days left until Christmas, checking our weight to determine how many pounds lost (or gained), using a measuring device to add ingredients to our food or to make coffee, or determining where to buy the cheapest gallon of gasoline, we are always measuring. We constantly take a measure of our world. In physics, measurement is fundamental; measurement is what physics is all about—why it exists. From the atomic scale to clusters of stars and galaxies, physicists must measure to probe the natural world.

In this chapter we begin where we should: at the beginning, with a presentation of fundamental math principles and operations. We can't even approach the fringe of comprehending physics unless we understand the basics of measurement, unit conversions, and basic math operations.

Measurement Units

The units most commonly used by physicists are based on the complicated English system of weights and measures. However, bench work is usually based on the metric system, or the International System of Units, due to the convenient relationship between milliliters (mL), cubic centimeters (cm^3), and grams (g).

The International System of Units (abbreviated SI—short for Système International d'Unités), is a modernized version of the metric system established by international agreement. The metric system of measurement was developed during the French Revolution and was first promoted in the United States in 1866. In 1902, proposed congressional legislation requiring the U.S. government to use the metric system exclusively was defeated by a single vote.

The most common measurement systems you see in physics are the centimeter-gram-second (CGS) and meter-kilogram-second (MKS) systems.

While we use both systems in this text, SI provides a logical and interconnected framework for all measurements in engineering, science, industry, and commerce. The metric system is much simpler to use than the existing English system, since all its units of measurement are divisible by 10.

Before listing the various conversion factors commonly used in physics it is important to describe the prefixes commonly used in the SI system. These prefixes are based on the power 10. For example, a kilo means 1000 grams, and a centimeter means one-hundredth of 1 meter. The twenty SI prefixes used to form decimal multiples and submultiples of SI units are given in table 1.1.

Note that the kilogram is the only SI unit with a prefix as part of its name and symbol. Because multiple prefixes may not be used, in the case of the kilogram the prefix names of table 1.1 are used with the unit name "gram" and the prefix symbols are used with the unit symbol "g." With this exception, any SI prefix may be used with any SI unit, including the degree Celsius and its symbol °C.

Example 1.1
10^{-6} kg = 1 mg (one milligram), but not 10^{-6} kg = 1 μkg (one microkilogram)

Example 1.2
Consider the height of the Washington Monument. We may write h_w = 169,000 mm = 16,900 cm = 169 m = 0.169 km using the millimeter (SI prefix "milli," symbol "m"), centimeter (SI prefix "centi," symbol "c"), or kilometer (SI prefix "kilo," symbol "k").

Table 1.1   SI prefixes

Factor	Name	Symbol
10^{24}	yotta	Y
10^{21}	zetta	Z
10^{18}	exa	E
10^{15}	peta	P
10^{12}	tera	T
10^{9}	giga	G
10^{6}	mega	M
10^{3}	kilo	k
10^{2}	hecto	h
10^{1}	deka	da
10^{-1}	deci	d
10^{-2}	centi	c
10^{-3}	milli	m
10^{-6}	micro	μ
10^{-9}	nano	n
10^{-12}	pico	p
10^{-15}	femto	f
10^{-18}	atto	a
10^{-21}	zepto	z
10^{-24}	yocto	y

Conversion Factors

Conversion factors refer to conversion of units between different units of measurement for the same quantity. It is important to note that the process of making a conversion cannot produce a more precise result than the original figure. Appropriate rounding of the results is normally performed after conversion. Conversion factors are given below in alphabetical order (table 1.2) and in order by unit category (table 1.3).

CONVERSION FACTORS: PRACTICAL EXAMPLES

Sometimes we have to convert between different units. Suppose that a 60-in. piece of pipe is attached to an existing 6-ft piece of pipe. Joined together, how long are they? Obviously, we cannot find the answer to this question by adding 60 to 6. Why? Because the two lengths are given in different units. Before we can add the two lengths, we must convert one of them to the units of the other. Then, when we have two lengths in the same units, we can add them.

To perform this conversion, we need a *conversion factor*. In this case, we have to know how many inches make up a foot—that is, 12 in. equal 1 ft. Knowing this, we can perform the calculation in two steps:

Step 1: 60 in. is really 60/12 = 5 ft
Step 2: 5 ft + 6 ft = 11 ft

From the example, we see that a conversion factor changes known quantities in one unit of measure to an equivalent quantity in another unit of measure.

In making the conversion from one unit to another, we must know two things:

• The exact number that relates the two units
• Whether to multiply or divide by that number

When making conversions, confusion over whether to multiply or divide is common; however, the number that relates the two units is usually known and thus is not a problem. Understanding the proper methodology—the "mechanics"—to use for various operations requires practice and common sense.

Along with using the proper "mechanics" (and practice and common sense) in making conversions, probably the easiest and fastest method of converting units is to use a conversion table. The simplest conversion requires that the measurement be multiplied or divided by a constant value. For instance, if the depth of wet cement in a form is 0.85 ft, multiplying by 12 in./ft converts the measured depth to inches (10.2 in.). Likewise, if the depth of the cement in the form is measured as 16 in., dividing by 12 in./ft converts the depth measurement to feet (1.33 ft).

Table 1.2   Alphabetical listing of conversion factors

Factors	*Metric (SI) or English conversions*
1 atm (atmosphere) =	1.013 bars
	10.133 newtons/cm^2 (newtons/square centimeter)
	33.90 ft of H_2O (feet of water)
	101.325 kp (kilopascals)
	1,013.25 mg (millibars)
	13.70 psia (pounds/square inch—absolute)
	760 torr
	760 mm Hg (millimeters of mercury)
1 bar =	0.987 atm (atmospheres)
	1 $\times$ 10^6 dynes/cm^2 (dynes/square centimeter)
	33.45 ft of H_2O (feet of water)
	1 $\times$ 10^5 pascals (nt/m^2) (newtons/square meter)
	750.06 torr
	750.06 mm Hg (millimeters of mercury)
1 Bq (becquerel) =	1 radioactive disintegration/second
	2.7 $\times$ 10^{-11} Ci (curie)
	2.7 $\times$ 10^{-8} mCi (millicurie)
1 Btu (British thermal unit) =	252 cal (calories)
	1,055.06 j (joules)
	10.41 liter-atmosphere
	0.293 watt-hours
1 cal (calorie) =	3.97 $\times$ 10^{-3} Btu (British thermal units)
	4.18 j (joules)
	0.0413 liter-atmospheres
	1.163 $\times$ 10^{-3} watt-hours
1 cc (cubic centimeter) =	3.53 $\times$ 10^{-5} ft^3 (cubic feet)
	0.061 in^3 (cubic inches)
	2.64 $\times$ 10^{-4} gal (gallons)
	0.001 ℓ (liters)
	1 ml (milliliters)
1 Ci (curie) =	3.7 (10^{10} radioactive disintegrations/second
	3.7 $\times$ 10^{10} Bq (becquerel)
	1,000 mCi (millicurie)
1 cm (centimeter) =	0.0328 ft (feet)
	0.394 in (inches)
	10,000 microns (micrometers)
	100,000,000 Å = 10^8 Å (Ångstroms)
1 cm^2 (square centimeter) =	1.076 $\times$ 10^{-3} ft^2 (square feet)
	0.155 in^2 (square inches)
	1 $\times$ 10^{-4} m^2 (square meters)

Table 1.2   (Continued)

Factors	Metric (SI) or English conversions
1 day =	24 hrs (hours) 1,440 min (minutes) 86,400 sec (seconds) 0.143 weeks 2.738×10^{-3} yrs (years)
1°C (expressed as an interval) =	1.8°F = (9/5)°F (degrees Fahrenheit) 1.8°R (degrees Rankine) 1.0 K (degrees Kelvin)
°C (degree Celsius) =	((5/9)(°F − 32°))
1°F (expressed as an interval) =	0.556°C = (5/9)°C (degrees Celsius) 1.0°R (degrees Rankine) 0.556 K (degrees Kelvin)
°F (degree Fahrenheit) =	((9/5)(°C) + 32°)
1 dyne =	1×10^{-5} nt (newton)
1 erg =	1 dyne-centimeter 1×10^{-7} j (joules) 2.78×10^{-11} watt-hours
1 ev (electron volt) =	1.602×10^{-12} ergs 1.602×10^{-19} j (joules)
1 fps (feet/second) =	1.097 kmph (kilometers/hour) 0.305 mps (meters/second) 0.01136 mph (miles/hour)
1 ft (foot) =	30.48 cm (centimeters) 12 in (inches) 0.3048 m (meters) 1.65×10^{-4} NM (nautical miles) 1.89×10^{-4} mi (statute miles)
1 ft² (square foot) =	2.296×10^{-5} acres 9.296 cm² (square centimeters) 144 in² (square inches) 0.0929 m² (square meters)
1 ft³ (cubic foot) =	28.317 cc (cubic centimeters) 1,728 in³ (cubic inches) 0.0283 m³ (cubic meters) 7.48 gal (gallons) 28.32 ℓ (liters) 29.92 qts (quarts)

Table 1.2   (Continued)

Factors	Metric (SI) or English conversions
1 gal (gallon) =	3,785 cc (cubic centimeters) 0.134 ft³ (cubic feet) 231 in³ (cubic inches) 3.785 ℓ (liters)
1 gm (gram)	0.001 kg (kilogram) 1,000 mg (milligrams) $1,000,000 \ ng = 10^6 \ ng$ (nanograms) 2.205×10^{-3} lbs (pounds)
1 gm/cc (gram/cubic centimeter) =	62.43 lbs/ft³ (pounds/cubic foot) 0.0361 lbs/in³ (pounds/cubic inch) 8.345 lbs/gal (pounds/gallon)
1 Gy (gray) =	1 j/kg (joules/kilogram) 100 rad 1 Sv (sievert) (unless modified through division by an appropriate factor, such as Q and/or N)
1 hp (horsepower) =	745.7 j/sec (joules/sec)
1 hr (hour) =	0.0417 days 60 min (minutes) 3,600 sec (seconds) 5.95×10^{-3} weeks 1.14×10^{-4} yrs (years)
1 in (inch) =	2.54 cm (centimeters) 1,000 mils
1 inch of water =	1.86 mm Hg (millimeters of mercury) 249.09 pascals 0.0361 psi (lbs/in²)
1 in³	16.39 cc (cubic centimeters) 16.39 ml (milliliters) 5.79×10^{-4} ft³ (cubic feet) 1.64×10^{-5} m³ (cubic meters) $4.33 \ (10^{-3}$ gal (gallons) 0.0164 ℓ (liters) 0.55 fl oz (fluid ounces)
1 j (joule) =	9.48×10^{-4} Btu (British thermal units) 0.239 cal (calories) $10,000,000 \ ergs = 1 \times 10^7$ ergs 9.87×10^{-3} liter-atmospheres 1.0 nt-m (newton-meters)

Table 1.2   (Continued)

Factors	Metric (SI) or English conversions
1 kcal (kilocalorie) =	3.97 Btu (British thermal units) 1,000 cal (calories) 4,186.8 j (joules)
1 kg (kilogram) =	1,000 gms (grams) 2,205 lbs (pounds)
1 km (kilometer) =	3,280 ft (feet) 0.54 NM (nautical miles) 0.6214 mi (statute miles)
1 kw (kilowatt) =	56.87 Btu/min (British thermal units/minute) 1.341 hp (horsepower) 1,000 j/sec (kilocalories)
1 kw-hr (kilowatt-hour) =	3,412.14 Btu (British thermal units) 3.6×10^6 j (joules) 859.8 kcal (kilocalories)
1 ℓ (liter) =	1,000 cc (cubic centimeters) 1 dm^3 (cubic decimeters) 0.0353 ft^3 (cubic feet) 61.02 in^3 (cubic inches) 0.264 gal (gallons) 1,000 ml (milliliters) 1.057 qts (quarts)
1 lb (pound) =	453.59 gms (grams) 16 oz (ounces)
l lbs/ft^3 (pounds/cubic foot) =	16.02 gms/l (grams/liter)
1 lbs/in^3 (pounds/cubic inch) =	27.68 gms/cc (grams/cubic centimeter) 1,728 lbs/ft^3 (pounds/cubic feet)
1 m (meter) =	1×10^{10} Å (Ångstroms) 100 cm (centimeters) 3.28 ft (feet) 39.37 in (inches) 1×10^{-3} km (kilometers) 1,000 mm (millimeters) 1,000,000 $\mu = 1 \times 10^6$ μ (micrometers) 1×10^9 nm (nanometers)
1 m^2 (square meter) =	10.76 ft^2 (square feet) 1,550 in^2 (square inches)

Table 1.2   (Continued)

Factors	Metric (SI) or English conversions
1 m³ (cubic meter) =	1,000,000 cc = 10^6 cc (cubic centimeters) 33.32 ft³ (cubic feet) 61,023 in³ (cubic inches) 264.17 gal (gallons) 1,000 ℓ (liters)
1 mi (statute mile) =	5,280 ft (feet) 1.609 km (kilometers) 1,609.3 m (meters) 0.869 NM (nautical miles) 1,760 yds (yards)
1 mi² (square mile) =	640 acres 2.79×10^7 ft² (square feet) 2.59×10^6 m² (square meters)
1 mCi (millicurie) =	0.001 Ci (curie) 3.7×10^{10} radioactive disintegrations/second 3.7×10^{10} Bq (becquerel)
1 min (minute) =	6.94×10^{-4} days 0.0167 hrs (hours) 60 sec (seconds) 9.92×10^{-5} weeks 1.90×10^{-6} yrs (years)
1 mm Hg (mm of mercury) =	1.316×10^{-3} atm (atmosphere) 0.535 in H_2O (inches of water) 1.33 mb (millibars) 133.32 pascals 1 torr 0.0193 psia (pounds/square inch—absolute)
1 mph (mile/hour) =	88 fpm (feet/minute) 1.61 kmph (kilometers/hour) 0.447 mps (meters/second)
1 mps (meters/second) =	196.9 fpm (feet/minute) 3.6 kmph (kilometers/hour) 2.237 mph (miles/hour)
1 NM (nautical mile) =	6,076.1 ft (feet) 1.852 km (kilometers) 1.15 mi (statute miles) 2,025.4 yds (yards)
1 nt (newton) =	1×10^5 dynes
1 nt-m (newton-meter) =	1.00 j (joules) 2.78×10^{-4} watt-hours

Table 1.2   (Continued)

Factors	Metric (SI) or English conversions
1 pascal =	9.87×10^{-6} atm (atmospheres) 4.015×10^{-3} in H_2O (inches of water) 0.01 mb (millibars) 7.5×10^{-3} mm Hg (milliliters of mercury)
1 ppm (parts/million—volume) =	1.00 ml/m³ (milliliters/cubic meter)
1 ppm (parts/million—weight) =	1.00 mg/kg (milligrams/kilograms)
1 psi (pounds/square inch) =	0.068 atm (atmospheres) 27.67 in H_2O (inches of water) 68.85 mb (millibars) 51.71 mm Hg (millimeters of mercury) 6,894.76 pascals
1 qt (quart) =	946.4 cc (cubic centimeters) 57.75 in³ (cubic inches) 0.946 ℓ (liters)
1 rad =	100 ergs/gm (ergs/gram) 0.01 Gy (gray) 1 rem (unless modified through division by an appropriate factor, such as Q and/or N)
1 rem	1 rad (unless modified through division by an appropriate factor, such as Q and/or N)
1 Sv (sievert) =	1 Gy (gray) (unless modified through division by an appropriate factor, such as Q and/or N)
1 torr =	1.33 mb (millibars)
1 watt =	3.41 Btu/hr (British thermal units/hour) 1.341×10^{-3} hp (horsepower) 52.18 j/sec (joules/second)
1 watt-hour =	3.412 Btu (British thermal unit) 859.8 cal (calories) 3,600 j (joules) 35.53 liter-atmosphere
1 week =	7 days 168 hrs (hours) 10,080 min (minutes) 6.048×10^5 sec (seconds) 0.0192 yrs (years)

Table 1.2   (Continued)

Factors	Metric (SI) or English conversions
1 yd³ (cubic yard) =	201.97 gal (gallons)
	764.55 ℓ (liters)
1 yr (year) =	365.25 days
	8,766 hrs (hours)
	5.26×10^5 min (minutes)
	3.16×10^7 sec (seconds)
	52.18 weeks

Table 1.3   Conversion factors by unit category

Units of Length

1 cm (centimeter) =	0.0328 ft (feet)
	0.394 in (inches)
	10,000 microns (micrometers)
	100,000,000 Å = 10^8 Å (Ångstroms)
1 ft (foot) =	30.48 cm (centimeters)
	12 in (inches)
	0.3048 m (meters)
	1.65×10^{-4} NM (nautical miles)
	1.89×10^{-4} mi (statute miles)
1 in (inch) =	2.54 cm (centimeters)
	1,000 mils
1 km (kilometer) =	3,280.8 ft (feet)
	0.54 NM (nautical miles)
	0.6214 mi (statute miles)
1 m (meter) =	1×10^{10} Å (Ångstroms)
	100 cm (centimeters)
	3.28 ft (feet)
	39.37 in (inches)
	1×10^{-3} km (kilometers)
	1,000 mm (millimeters)
	1,000,000 μ = 1×10^6 μ (micrometers)
	1×10^9 nm (nanometers)
1 mi (statute mile) =	5,280 ft (feet)
	1.609 km (kilometers)
	1.690.3 m (meters)
	0.869 NM (nautical miles)
	1,760 yd (yards)

Table 1.3   (Continued)

| 1 NM (nautical mile) = | 6,076.1 ft (feet)
1.852 km (kilometers)
1.15 mi (statute miles)
2.025.4 yd (yards) |

Units of Area

1 cm² (square centimeter) =	1.076×10^{-3} ft² (square feet) 0.155 in² (square inches) 1×10^{-4} m² (square meters)
1 ft² (square foot) =	2.296×10^{-5} acres 929.03 cm² (square centimeters) 144 in² (square inches) 0.0929 m² (square meters)
1 m² (square meter) =	10.76 ft² (square feet) 1,550 in² (square inches)
1 mi² (square mile) =	640 acres 2.79×10^{7} ft² (square feet) 2.59×10^{6} m² (square meters)

Units of Volume

1 cc (cubic centimeter) =	3.53×10^{-5} ft³ (cubic feet) 0.061 in³ (cubic inches) 2.64×10^{-4} gal (gallons) 0.001 ℓ (liters) 1.00 ml (milliliters)
1 ft³ (cubic foot) =	28,317 cc (cubic centimeters) 1,728 in³ (cubic inches) 0.0283 m³ (cubic meters) 7.48 gal (gallons) 28.32 ℓ (liters) 29.92 qt (quarts)
1 gal (gallon) =	3,785 cc (cubic centimeters) 0.134 ft³ (cubic feet) 231 in³ (cubic inches) 3.785 ℓ (liters)
1 in³ (cubic inch) =	16.39 cc (cubic centimeters) 16.39 ml (milliliters) 5.79×10^{-4} ft³ (cubic feet) 1.64×10^{-5} m³ (cubic meters) 4.33×10^{-3} gal (gallons) 0.0164 ℓ (liters) 0.55 fl oz (fluid ounces)

Table 1.3   (Continued)

1 ℓ (liter) =	1,000 cc (cubic centimeters)
	1 dm^3 (cubic decimeters)
	0.0353 ft^3 (cubic feet)
	61.02 in^3 (cubic inches)
	0.264 gal (gallons)
	1,000 (milliliters)
	1.057 qt (quarts)
1 m^3 (cubic meter) =	1,000,000 cc = 10^6 cc (cubic centimeters)
	35.31 ft^3 (cubic feet)
	61,023 in^3 (cubic inches)
	264.17 gal (gallons)
	1,000 ℓ (liters)
1 qt (quart) =	946.4 cc (cubic centimeters)
	57.75 in^3 (cubic inches)
	0.946 ℓ (liters)
1 yd^3 (cubic yard) =	201.97 gal (gallons)
	764.55 ℓ (liters)

Units of Mass

1 gm (gram) =	0.001 kg (kilograms)
	1,000 mg (milligrams)
	1,000,000 mg = 10^6 ng (nanograms)
	2.205 × 10^{-3} lb (pounds)
1 kg (kilogram) =	1,000 gm (grams)
	2.205 lb (pounds)
1 lb (pound) =	453.59 gm (grams)
	16 oz (ounces)

Units of Time

1 day =	24 hr (hours)
	1440 min (minutes)
	86,400 sec (seconds)
	0.143 weeks
	2.738 × 10^{-3} yr (years)
1 hr (hour) =	0.0417 days
	60 min (minutes)
	3,600 sec (seconds)
	5.95 × 10^{-3} weeks
	1.14 × 10^{-4} yr (years)
1 min (minutes) =	6.94 × 10^{-4} days
	0.0167 hr (hours)
	60 sec (seconds)
	9.92 × 10^{-5} weeks
	1.90 × 10^{-6} yr (years)

Table 1.3   (Continued)

1 week =	7 days
	168 hr (hours)
	10,080 min (minutes)
	6.048×10^5 sec (seconds)
	0.0192 yr (years)
1 yr (year) =	365.25 days
	8,766 hr (hours)
	5.26×10^5 min (minutes)
	3.16×10^7 sec (seconds)
	52.18 weeks

Units of Temperature

°C (degrees Celsius) =	$((5/9)(°F - 32°))$
1°C (expressed as an interval) =	$1.8°F = (9/5)°F$ (degrees Fahrenheit)
	1.8°R (degrees Rankine)
	1.0 K (degrees Kelvin)
°F (degree Fahrenheit) =	$((9/5)(°C) + 32)$
1°F (expressed as an interval) =	$0.556°C = (5/9)°C$ (degrees Celsius)
	1.0°R (degrees Rankine)
	0.556 K (degrees Kelvin)

Units of Force

1 dyne =	1×10^{-5} nt (newtons)
1 nt (newton) =	1×10^5 dynes

Units of Work or Energy

1 Btu (British thermal unit) =	252 cal (calories)
	1,055.06 j (joules)
	10.41 liter-atmospheres
	0.293 watt-hours
1 cal (calories) =	3.97×10^{-3} Btu (British thermal units)
	4.18 j (joules)
	0.0413 liter-atmospheres
	1.163×10^{-3} watt-hours
1 erg =	1 dyne-centimeter
	1×10^{-7} j (joules)
	2.78×10^{-11} watt-hours
1 ev (electron volt) =	1.602×10^{-12} ergs
	1.602×10^{-19} j (joules)

Table 1.3   (Continued)

1 j (joule) =	9.48×10^{-4} Btu (British thermal units)
	0.239 cal (calories)
	10,000,000 ergs = 1×10^7 ergs
	9.87×10^{-3} liter-atmospheres
	1.00 nt-m (newton-meters)
1 kcal (kilocalorie) =	3.97 Btu (British thermal units)
	1,000 cal (calories)
	4,186.8 j (joules)
1 kw-hr (kilowatt-hour) =	3,412.14 Btu (British thermal units)
	3.6×10^6 j (joules)
	859.8 kcal (kilocalories)
1 nt-m (newton-meter) =	1.00 j (joules)
	2.78×10^{-4} watt-hours
1 watt-hour =	3.412 Btu (British thermal units)
	859.8 cal (calories)
	3,600 j (joules)
	35.53 liter-atmospheres

Units of Power

1 hp (horsepower) =	745.7 j/sec (joules/sec)
1 kw (kilowatt) =	56.87 Btu/min (British thermal units/minute)
	1.341 hp (horsepower)
	1,000 j/sec (joules/sec)
1 watt =	3.41 Btu/hr (British thermal units/hour)
	1.341×10^{-3} hp (horsepower)
	1.00 j/sec (joules/second)

Units of Pressure

1 atm (atmosphere) =	1.013 bars
	10.133 newtons/cm^2 (newtons/square centimeters)
	33.90 ft of H_2O (feet of water)
	101.325 kp (kilopascals)
	14.70 psia (pounds/square inch—absolute)
	760 torr
	760 mm Hg (millimeters of mercury)
1 bar =	0.987 atm (atmospheres)
	1×10^6 dynes/cm^2 (dynes/square centimeter)
	33.45 ft of H_2O (feet of water)
	1×10^5 pascals (nt/m^2) (newtons/square meter)
	750.06 torr
	750.06 mm Hg (millimeters of mercury)

Table 1.3   (Continued)

1 inch of water =	1.86 mm Hg (millimeters of mercury) 249.09 pascals 0.0361 psi (lbs/in²)
1 mm Hg (millimeter of merc.) =	1.316×10^{-3} atm (atmospheres) 0.535 in H_2O (inches of water) 1.33 mb (millibars) 133.32 pascals 1 torr 0.0193 psia (pounds/square inch—absolute)
1 pascal =	9.87×10^{-6} atm (atmospheres) 4.015×10^{-3} in H_2O (inches of water) 0.01 mb (millibars) 7.5×10^{-3} mm Hg (millimeters of mercury)
1 psi (pounds/square inch) =	0.068 atm (atmospheres) 27.67 in H_2O (inches of water) 68.85 mb (millibars) 51.71 mm Hg (millimeters of mercury) 6,894.76 pascals
1 torr =	1.33 mb (millibars)

Units of Velocity or Speed

1 fps (feet/second) =	1.097 kmph (kilometers/hour) 0.305 mps (meters/second) 0.01136 mph (miles/hours)
1 mph (miles/hour) =	88 fpm (feet/minute) 1.61 kmph (kilometers/hour) 0.447 mps (meters/second)
1 mps (meters/second) =	196.9 fpm (feet/minute) 3.6 kmph (kilometers/hour) 2.237 mph (miles/hour)

Units of Density

1 gm/cc (gram/cubic cent.) =	62.43 lb/ft³ (pounds/cubic foot) 0.0361 lb/in³ (pounds/cubic inch) 8.345 lb/gal (pounds/gallon)
1 lb/ft³ (pound/cubic foot) =	16.02 gm/l (grams/liter)
1 lb/in² (pound/cubic inch) =	27.68 gm/cc (grams/cubic centimeter) 1.728 lb/ft³ (pounds/cubic foot)

Units of Concentration

1 ppm (parts/million—volume) =	1.00 ml/m³ (milliliters/cubic meter)

Table 1.3   (Continued)

1 ppm (parts/million—weight) =	1.00 mg/kg (milligrams/kilograms)

Radiation and Dose-Related Units

1 Bq (becquerel) =	1 radioactive disintegration/second 2.7×10^{-11} Ci (curie) 2.7×10^{-8} mCi (millicurie)
1 Ci (curie) =	3.7×10^{10} radioactive disintegration/second 3.7×10^{10} Bq (becquerel) 1,000 mCi (millicurie)
1 Gy (gray) =	1 j/kg (joule/kilogram) 100 rad 1 Sv (sievert) (unless modified through division by an appropriate factor, such as Q and/or N)
1 mCi (millicurie) =	0.001 Ci (curie) 3.7×10^{10} radioactive disintegrations/second 3.7×10^{10} Bq (becquerel)
1 rad =	100 ergs/gm (ergs/gm) 0.01 Gy (gray) 1 rem (unless modified through division by an appropriate factor, such as Q and/or N)
1 rem =	1 rad (unless modified through division by an appropriate factor, such as Q and/or N)
1 Sv (sievert) =	1 Gy (gray) (unless modified through division by an appropriate factor, such as Q and/or N)

Weight, Concentration, and Flow

Using table 1.4 to convert from one unit expression to another and vice versa is good practice. However, in making conversions to solve process computations in water treatment operations, for example, we must be familiar with conversion calculations based upon a relationship between weight, flow or volume, and concentration. The basic relationship is

$$\text{Weight} = \text{Concentration} \times \text{Flow or Volume} \times \text{Factor} \qquad (1.1)$$

Table 1.5 summarizes weight, volume, and concentration calculations. With practice, many of these calculations become second nature to users.

The following conversion factors are used extensively in physics—in water/waste-

Table 1.4   Conversion table

To convert	Multiply by	To get
Feet	12	Inches
Yards	3	Feet
Yards	36	Inches
Inches	2.54	Centimeters
Meters	3.3	Feet
Meters	100	Centimeters
Meters	1,000	Millimeters
Square Yards	9	Square Feet
Square Feet	144	Square Inches
Acres	43,560	Square Feet
Cubic Yards	27	Cubic Feet
Cubic Feet	1,728	Cubic Inches
Cubic Feet (water)	7.48	Gallons
Cubic Feet (water)	62.4	Pounds
Acre-Feet	43,560	Cubic Feet
Gallons (water)	8.34	Pounds
Gallons (water)	3.785	Liters
Gallons (water)	3,785	Milliliters
Gallons (water)	3,785	Cubic Centimeters
Gallons (water)	3,785	Grams
Liters	1,000	Milliliters
Days	24	Hours
Days	1,440	Minutes
Days	86,400	Seconds
Million Gallons/Day	1,000,000	Gallons/Day
Million Gallons/Day	1.55	Cubic Feet/Second
Million Gallons/Day	3.069	Acre-Feet/Day
Million Gallons/Day	36.8	Acre-Inches/Day
Million Gallons/Day	3,785	Cubic Meters/Day
Gallons/Minute	1,440	Gallons/Day
Gallons/Minute	63.08	Liters/Minute
Pounds	454	Grams
Grams	1,000	Milligrams
Pressure, psi	2.31	Head, Feet (Water)
Horsepower	33,000	Foot-Pounds/Minute
Horsepower	0.746	Kilowatts
To get	Divide by	To convert

water calculations. Water calculations are used frequently in this text because they are germane to the subject and illustrative of commonly used math operations.

- 7.48 gallons per ft³
- 3.785 liters per gallon
- 454 grams per pound
- 1,000 mL per liter
- 1,000 mg per gram
- 1 ft³/sec (cfs) = 0.6465 MGD

Table 1.5   Weight, volume, and concentration calculations

To calculate	Formula
Pounds	Concentration, mg/L × Tank Vol. MG × 8.34 lb/MG/mg/L
Pounds/Day	Concentration, mg/L × Flow, MGD × 8.34 lb/MG/mg/L
Million Gallons/Day	$\dfrac{\text{Quantity, lb/day}}{(\text{Conc., mg/L} \times 8.34 \text{ lb/mg/L/MG})}$
Milligrams/Liter	$\dfrac{\text{Quantity, lb}}{(\text{Tank Volume, MG} \times 8.34 \text{ lb/mg/L/MG})}$
Kilograms/Liter	Conc., mg/L × Volume, MG × 3.785 lb/MG/mg/L
Kilograms/Day	Conc., mg/L × Flow, MGD × 3.785 lb/MG/mg/L
Pounds/Dry Ton	Conc. Mg/kg × 0.002 lb/d.t./mg/kg

✔ **Key Point:** Density (also called specific weight) is mass per unit volume, and may be registered as lb/cu ft, lb/gal, grams/mL, or grams/cu meter. If we take a fixed-volume container, fill it with a fluid, and weigh it, we can determine the density of the fluid (after subtracting the weight of the container).

- 8.34 pounds per gallon (water)—(density = 8.34 lb/gal)
- one milliliter of water weighs 1 gram—(density = 1 gram/mL)
- 62.4 pounds per ft³ (water)—(density = 8.34 lb/gal)
- 8.34 lb/gal = mg/L (converts dosage in mg/L into lb/day/MGD)
  - example: 1 mg/L × 10 MGD × 8.3 = 83.4 lb/day
- 1 psi = 2.31 feet of water (head)
- 1 foot head = 0.433 psi
- °F = 9/5 (°C + 32)
- °C = 5/9 (°F − 32)

✔ **Note:** Use tables 1.4 and 1.5 to make the conversions indicated in the following example problems. Other conversions are presented in appropriate sections of the text.

Example 1.3
Convert cubic feet to gallons.

$$\text{Gallons} = \text{Cubic Feet, ft}^3 \times \text{gal/ft}^3$$

Sample Problem
How many gallons of biosolids (i.e., wastewater sludge) can be pumped to a digester that has 3,600 cubic feet of volume available?

$$\text{Gallons} = 3,600 \text{ ft}^3 \times 7.48 \text{ gal/ft}^3 = 26,928 \text{ gal}$$

Example 1.4
Convert gallons to cubic feet.

$$\text{Cubic Feet} = \frac{\text{gal}}{7.48 \text{ gal/ft}^3}$$

Sample Problem

How many cubic feet of biosolids are removed when 18,200 gallons are withdrawn?

$$\text{Cubic Feet} = \frac{18,200 \text{ gal}}{7.48 \text{ gal/ft}^3} = 2,433 \text{ ft}^3$$

Example 1.5

Convert gallons to pounds.

$$\text{Pounds, lb} = \text{Gal} \times 8.34 \text{ lb/gal}$$

Sample Problem

If 1,650 gallons of solids are removed from the primary settling tank, how many pounds of solids are removed?

$$\text{Pounds} = 1,650 \text{ gal} \times 8.34\text{/gal} = 13,761 \text{ lb}$$

Example 1.6

Convert pounds to gallons.

$$\text{Gallons} = \frac{\text{lb}}{8.34 \text{ lb/gal}}$$

Sample Problem

How many gallons of water are required to fill a tank that holds 7,540 pounds of water?

$$\text{Gallons} = \frac{7,540 \text{ lb}}{8.34 \text{ lb/gal}} = 904 \text{ gal}$$

Example 1.7

Convert milligrams/liter to pounds.

✔ **Key Point:** For plant operations, concentrations in milligrams per liter or parts per million determined by laboratory testing must be converted to quantities of pounds, kilograms, pounds per day, or kilograms per day.

$$\text{Pounds} = \text{Concentration, mg/L} \times \text{Volume, MG} \times 8.34 \text{ lb/mg/L/MG}$$

Sample Problem

The solids concentration in the aeration tank is 2,580 mg/L. The aeration tank volume is 0.95 MG. How many pounds of solids are in the tank?

Pounds = 2,580 mg/L × 0.95 MG × 8.34 lb/mg/L/MG = 20,441.3 lb

Example 1.8

Convert milligrams per liter to pounds per day.

Pounds/Day = Concentration, mg/L × Flow, MGD × 8.34 lb/mg/L/MG

Sample Problem

How many pounds of solids are discharged per day when the plant effluent flow rate is 4.75 MGD and the effluent solids concentration is 26 mg/L?

Pounds/Day = 26 mg/L × 4.75 MGD × 8.34 lb/mg/L/MG = 1,030 lb/day

Example 1.9

Convert milligrams per liter to kilograms per day.

Kg/Day = Concentration, mg/L × Volume, MG × 3.785 kg/mg/L/MG

Sample Problem

The effluent contains 26 mg/L of BOD_5. How many kilograms per day of BOD_5 are discharged when the effluent flow rate is 9.5 MGD?

Kg/Day = 26 mg/L × 9.5 MG × 3.785 kg/mg/L/MG = 935 kg/day

Example 1.10

Convert pounds to milligrams per liter.

$$\text{Concentration, mg/L} = \frac{\text{Quantity, lb}}{\text{Volume, MG} \times 8.34 \text{ lb/mg/L/MG}}$$

Sample Problem

The aeration tank contains 89,990 pounds of solids. The volume of the aeration tank is 4.45 MG. What is the concentration of solids in the aeration tank in mg/L?

$$\text{Concentration, mg/L} = \frac{89,990 \text{ lb}}{4.45 \text{ MG} \times 8.34 \text{ lb/mg/L/MG}} = 2,425 \text{ mg/L}$$

Example 1.11

Convert pounds per day to milligrams per liter.

$$\text{Concentration, mg/L} = \frac{\text{Quantity, lb/day}}{\text{Volume, MGD} \times 8.34 \text{ lb/mg/L/MG}}$$

Sample Problem

The disinfection process uses 4,820 pounds per day of chlorine to disinfect a flow of 25.2 MGD. What is the concentration of chlorine applied to the effluent?

$$\text{Concentration, mg/L} = \frac{4,820}{25.2 \text{ MGD} \times 8.34 \text{ lb/mg/L/MG}} = 22.9 \text{ mg/L}$$

Example 1.12
Convert pounds to flow in million gallons per day.

$$\text{Flow} = \frac{\text{Quantity, lb/day}}{\text{Concentration, mg/L} \times 8.34 \text{ lb/mg/L/MG}}$$

Sample Problem
Each day, 9,640 pounds of solids must be removed by the activated biosolids process. The waste activated biosolids concentration is 7,699 mg/L. How many million gallons per day of waste activated biosolids must be removed?

$$\text{Flow} = \frac{9,640 \text{ lb}}{7,699 \text{ mg/L} \times 8.34 \text{ lb/MG/mg/L}} = 0.15 \text{ MGD}$$

Example 1.13
Convert million gallons per day (MGD) to gallons per minute (gpm).

$$\text{Flow} = \frac{\text{Flow, MGD} \times 1,000,000 \text{ gal/MG}}{1,440 \text{ min/day}}$$

Sample Problem
The current flow rate is 5.55 MGD. What is the flow rate in gallons per minute?

$$\text{Flow} = \frac{5.55 \text{ MGD} \times 1,000,000 \text{ gal/MG}}{1,440 \text{ min/day}} = 3,854 \text{ gpm}$$

Example 1.14
Convert million gallons per day (MGD) to gallons per day (gpd).

$$\text{Flow} = \text{Flow, MGD} \times 1,000,000 \text{ gal/MG}$$

Sample Problem
The influent meter reads 28.8 MGD. What is the current flow rate in gallons per day?

$$\text{Flow} = 28.8 \text{ MGD} \times 1,000,000 \text{ gal/MG} = 28,800,000 \text{ gpd}$$

Example 1.15
Convert million gallons per day (MGD) to cubic feet per second (cfs).

$$\text{Flow, cfs} = \text{Flow, MGD} \times 1.55 \text{ cfs/MGD}$$

Sample Problem
The flow rate entering the grit channel is 2.89 MGD. What is the flow rate in cubic feet per second?

$$\text{Flow} = 2.89 \text{ MGD} \times 1.55 \text{ cfs/MGD} = 4.48 \text{ cfs}$$

Example 1.16
Convert gallons per minute (gpm) to million gallons per day (MGD).

$$\text{Flow, MGD} = \frac{\text{Flow, gpm} \times 1,440 \text{ min/day}}{1,000,000 \text{ gal/MG}}$$

Sample Problem
The flow meter indicates that the current flow rate is 1,469 gpm. What is the flow rate in MGD?

$$\text{Flow, MGD} = \frac{1,469 \text{ gpm} \times 1,440 \text{ min/day}}{1,000,000 \text{ gal/MG}} = 2.12 \text{ MGD (rounded)}$$

Example 1.17
Convert gallons per day (gpd) to million gallons per day (MGD).

$$\text{Flow, MGD} = \frac{\text{Flow, gal/day}}{1,000,000 \text{ gal/MG}}$$

Sample Problem
The totalizing flow meter indicates that 33,444,950 gallons of wastewater have entered the plant in the past 24 hours. What is the flow rate in MGD?

$$\text{Flow} = \frac{33,444,950 \text{ gal/day}}{1,000,000 \text{ gal/MG}} = 33.44 \text{ MGD}$$

Example 1.18
Convert flow in cubic feet per second (cfs) to million gallons per day (MGD).

$$\text{Flow, MGD} = \frac{\text{Flow, cfs}}{1.55 \text{ cfs/MG}}$$

Sample Problem
The flow in a channel is determined to be 3.89 cubic feet per second (cfs). What is the flow rate in million gallons per day (MGD)?

$$\text{Flow, MGD} = \frac{3.89 \text{ cfs}}{1.55 \text{ cfs/MG}} = 2.5 \text{ MGD}$$

Example 1.19
Problem
The water in a tank weighs 675 lb. How many gallons does it hold?

Solution
Water weighs 8.34 lbs/gal. Therefore:

$$\frac{675 \text{ lb}}{8.34 \text{ lb/gal}} = 80.9 \text{ gallons}$$

Example 1.20
Problem
A liquid chemical weighs 62 lb/cu ft. How much does a 5-gallon can of it weigh?

Solution
Solve for specific gravity; get lb/gal; multiply by 5.

$$\text{Specific Gravity} = \frac{\text{wt. channel}}{\text{wt. water}}$$

$$\frac{62 \text{ lb/cu ft}}{62.4 \text{ lb/cu ft}} = .99$$

$$\text{Specific Gravity} = \frac{\text{wt. chemical}}{\text{wt. water}}$$

$$.99 = \frac{\text{wt. chemical}}{8.34 \text{ lb/gal}}$$

$$8.26 \text{ lb/gal} = \text{wt. chemical}$$

$$8.26 \text{ lb/gal} \times 5 \text{ gal} = 41.3 \text{ lb}$$

Example 1.21
Problem
A wooden piling with a diameter of 16 in. and a length of 16 ft weighs 50 lb/cu ft. If it is inserted vertically into a body of water, what vertical force is required to hold it below the water surface?

Solution
If this piling had the same weight as water, it would rest just barely submerged. Find the difference between its weight and that of the same volume of water. That is the weight needed to keep it down.

$$\begin{array}{r} 62.4 \text{ lb/cu ft (water)} \\ - 50.0 \text{ lb/cu ft (piling)} \\ \hline 12.4 \text{ lb/cu ft difference} \end{array}$$

Volume of piling = .785 × 1.33² × 16 ft = 22.22 cu ft

12.4 lb/cu ft × 22.22 cu ft = 275.5 lb (needed to hold it below water surface)

Example 1.22
Problem
A liquid chemical with a specific gravity (SG) of 1.22 is pumped at a rate of 40 gpm. How many pounds per day are being delivered by the pump?

Solution
Solve for pounds pumped per minute; change to lb/day.
8.34 lb/gal water × 1.22 SG liquid chemical = 10.2 lb/gal liquid
40 gal/min × 10.2 lb/gal = 408 lb/min
408 lb/min × 1,440 min/day = 587,520 lb/day.

Example 1.23
Problem
A cinder block weighs 70 lb in air. When immersed in water, it weighs 40 lb. What is the volume and specific gravity of the cinder block?

Solution
The cinder block displaces 30 lb of water; solve for cu ft of water displaced (equivalent to volume of cinder block).

$$\frac{30 \text{ lb water displaced}}{62.4 \text{ lb/cu ft}} = .48 \text{ cu ft water displaced}$$

Cinder block volume = .48 cu ft; this weighs 70 lb.

$$\frac{70 \text{ lb}}{.48 \text{ cu ft}} = 145.8 \text{ lb/cu ft density of cinder block}$$

$$\text{Specific Gravity} = \frac{\text{density of cinder block}}{\text{density of water}}$$

$$= \frac{145.8 \text{ lb/cu ft}}{62.4 \text{ lb/cu ft}}$$

$$= 2.34$$

TEMPERATURE CONVERSIONS

There are two commonly used methods used to make temperature conversions. We have already demonstrated the following method:

$$°C = 5/9 \ (°F - 32)$$
$$°F = 9/5 \ (°C) + 32$$

Example 1.24
Problem
At a temperature of 4°C, water is at its greatest density. What is the degree of Fahrenheit?

Solution

$$°F = (°C) \times 9/5 + 32$$
$$= 4 \times 9/5 + 32$$
$$= 7.2 + 32$$
$$= 39.2$$

However, the difficulty arises when one tries to recall these formulas from memory. Probably the easiest way to recall these important formulae is to remember three basic steps for both Fahrenheit and Celsius conversions:

Step 1: Add 40
Step 2: Multiply by the appropriate fraction (5/9 or 9/5)
Step 3: Subtract 40°

Obviously, the only variable in this method is the choice of 5/9 or 9/5 in the multiplication step. To make the proper choice, you must be familiar with the two scales. The freezing point of water is 32° on the Fahrenheit scale and 0° on the Celsius scale. The boiling point of water is 212° on the Fahrenheit scale and 100° on the Celsius scale.
What does all this mean?

✔ **Key Point:** Note, for example, that at the same temperature, higher numbers are associated with the Fahrenheit scale and lower numbers with the Celsius scale. This important relationship helps you decide whether to multiply by 5/9 or 9/5. Let's look at a few conversion problems to see how the three-step process works.

Example 1.25
Suppose that we wish to convert 240°F to Celsius. Using the three-step process, we proceed as follows:

Step 1: Add 40°.

$$240° + 40° = 280°$$

Step 2: We must multiply 280° by either 5/9 or 9/5. Because the conversion is to the Celsius scale, we will be moving to a number *smaller* than 280. Through reason and observation, obviously, if 280 were multiplied by 9/5, the result would be almost the same as multiplying by 2, which would double 280 rather than make it smaller. If we multiply by 5/9, the result will be about the same as multiplying by $\frac{1}{2}$, which would cut 280 in half. Because in this problem we wish to move to a smaller number, we should multiply by 5/9:

$$(5/9) \; (280°) \; = \; 156.0°C$$

Step 3: Now subtract 40°.

$$156.0°C \; - \; 40.0°C \; = \; 116.0°C$$

Therefore, 240°F $=$ 116.0°C.

Example 1.26
Convert 22°C to Fahrenheit.

(1) Step 1: Add 40°.

$$22° \; + \; 40° \; = \; 62°$$

(2) Step 2: Because we are converting from Celsius to Fahrenheit, we are moving from a smaller to a larger number, and 9/5 should be used in the multiplications:

$$(9/5) \; (62°) \; = \; 112°$$

(3) Step 3: Subtract 40°.

$$112° \; - \; 40° \; = \; 72°$$

Thus, 22°C $=$ 72°F.

Obviously, knowing how to make these temperature conversion calculations is useful. However, in practical *in situ* or non–*in situ* operations, you may wish to use a temperature conversion table.

CONVERSION FACTORS: AIR POLLUTION MEASUREMENTS

Physicists are involved in the science, engineering, and design of air pollution control equipment. Thus, sample air pollution control measurements are included in this text. The recommended units for reporting air pollutant emissions are commonly stated in metric system whole numbers. If possible, the reported units should be the same as those that are actually being measured. For example, weight should be recorded in grams; volume of air should be recorded in cubic meters. When the analytical system is calibrated in one unit, the emissions should also be reported in the units of the calibration standard. For example, if a gas chromatograph is calibrated with a 1-ppm standard of toluene in air, then the emissions monitored by the system should also be reported in ppm. Finally, if the emission standard is defined in a specific unit, the monitoring system should be selected to monitor in that unit.

The preferred reporting units for the following types of emissions should be:

- Nonmethane organic and volatile organic compound emissions   ppm, ppb
- Semi-volatile organic compound emissions $\mu g/m^3$, mg/m^3
- Particulate matter (TSP/PM-10) emissions $\mu g/m^3$
- Metal compound emissions ng/m^3

Conversion from ppm to μg/m³

Often, the physicist must be able to convert from ppm to $\mu g/m^3$. Following is an example of how one would perform that conversion using sulfur dioxide (SO_2) as the monitored constituent.

Example 1.27

The expression "parts per million" is without dimensions, i.e., no units of weight or volume are specifically designed. Using the format of other units, the expression may be written:

$$\frac{parts}{million\ parts}$$

"Parts" are not defined. If cubic centimeters replace parts, we obtain:

Similarly, we might write "pounds per million pounds," "tons per million tons," or "liters per million liters." In each expression, identical units of weight or volume appear in both the numerator and denominator and may be canceled out, leaving a dimensionless term.

An analog of parts per million is the more familiar term "percent." Percent can be written:

$$\frac{parts}{hundred\ parts}$$

To convert from part per million by volume, ppm, ($\mu L/L$), to $\mu g/m^3$ at EPA's standard temperature (25°C) and standard pressure (760 mm Hg), STP, it is necessary to know the molar volume at the given temperature and pressure and the molecular weight of the pollutant.

At 25°C and 760 mm Hg, one mole of any gas occupies 24.46 liters.

Problem

The atmospheric concentration of sulfur dioxide (SO_2) by volume was reported as 2.5 ppm. What is this concentration in micrograms (μg) per cubic meter (m^3) at 25°C and 760 mm Hg? What is the concentration in $\mu g/m^3$ at 37°C and 752 mm Hg?

✔ **Note:** The following example problem points out the need for reporting temperature and pressure when the results are present on a weight to volume basis.

Solution

Let parts per million equal $\mu L/L$; then 2.5 ppm $= 2.5\ \mu L/L$. The molar volume at 25°C and 760 mm Hg is 24.46 L, and the molecular weight of SO_2 is 64.1 g/mole.

Step 1: 25°C and 760 mm Hg

$$\frac{2.5\ \mu L}{L} \times \frac{1\ \mu mole}{24.46\ \mu L} \times \frac{64.1\ \mu g}{\mu mole} \times \frac{1,000\ L}{m^3} = \frac{6.6 \times 10^3\ \mu g}{m^3}\ \text{at STP}$$

Step 2: 37°C and 752 mm Hg

$$24.46\ \mu L \left(\frac{310°K}{298°D}\right)\left(\frac{760\ mm\ Hg}{752\ mm\ Hg}\right) = 25.72\ \mu L$$

$$\frac{2.5\ \mu L}{L} \times \frac{1\ \mu mole}{25.72\ \mu L} \times \frac{64.1\ \mu g}{\mu mole} \times \frac{1,000\ L}{m^3} \times \frac{6.2 \times 10^3\ \mu g}{m^3}\ \text{at 37°, 752 mm Hg}$$

Conversion Tables for Common Air Pollution Measurements

To assist the physicist in converting from one set of units to another, the following conversion factors for common air pollution measurements and other useful information are provided. The conversion tables provide factors for:

• Atmospheric gases
• Atmospheric pressure
• Gas velocity
• Concentration
• Atmospheric particulate matter

Following is a list of conversions from ppm to $\mu g/m^3$ (at 25°C and 760 mm Hg) for several common air pollutants:

ppm SO_2 × 2620 = $\mu g/m^3$ SO_2 (sulfur dioxide)
ppm CO × 1150 = $\mu g/m^3$ CO (carbon monoxide)
ppm CO_x × 1.15 = mg/m^3 CO (carbon dioxide)
ppm CO_2 × 1.8 = mg/m^3 CO_2 (carbon dioxide)
ppm NO × 1230 = $\mu g/m^3$ NO (nitrogen oxide)
ppm NO_2 × 1880 = $\mu g/m^3$ NO_2 (nitrogen dioxide)
ppm O_2 × 1960 = $\mu g/m^3$ O_3 (ozone)
ppm CH_4 × 655 = $\mu g/m^3$ CH_4 (methane)
ppm CH_4 × 655 = mg/m^3 CH_4 (methane)
ppm CH_3SH × 2000 = $\mu g/m^3$ CH_3SH (methyl mercaptan)
ppm C_3H_8 × 1800 = $\mu g/m^3$ C_3H_8 (propane)
ppm C_3H_8 × 1.8 = mg/m^3 C_3H_8 (propane)
ppm F- × 790 = $\mu g/m^3$ F- (fluoride)
ppm H_2S × 1400 = $\mu g/m^3$ H_2S (hydrogen sulfide)
ppm NH_3 × 696 = $\mu g/m^3$ NH_3 (ammonia)
ppm HCHO × 1230 = $\mu g/m^3$ HCHO (formaldehyde)

Tables 1.6 through 1.10 shows various conversion calculations.

SOIL TEST RESULTS CONVERSION FACTORS

Soil test results can be converted from parts per million (ppm) to pounds per acre by multiplying ppm by a conversion factor based on the depth to which the soil was sampled. Because a slice of soil 1 acre in area and 3 inches deep weighs approximately 1 million pounds, the conversion factors given in table 1.11 can be used.

Basic Math Operations

Most calculations required by physicists (as with many others) start with the basics, such as addition, subtraction, multiplication, division, and sequence of operations. Although many of the operations are fundamental tools within each physicist's tool-box, using these tools on a consistent basis is important in order to remain sharp in their use. Physicists should master basic math definitions and the formation of problems; daily practice requires calculation of percentages, averages, simple ratios; geometric dimensions, force, pressure, and head, as well as the use of dimensional analysis and advanced math operations.

BASIC MATH TERMINOLOGY AND DEFINITIONS

The following basic definitions will aid in understanding the material in this chapter.

- *Integer,* or an *integral number*: a whole number. Thus 1, 2, 3, 4, 5, 6, 7, 8, 9, 10, 11, and 12 are the first 12 positive integers.

Table 1.6   Atmospheric gases

To convert from	To	Multiply by
Milligram/cu m	Micrograms/cu m	1000.0
	Micrograms/liter	1.0
	ppm by volume (20°C)	24.04/M
	ppm by weight	0.8347
	Pounds/cu ft	62.43×10^{-9}
Micrograms/cu ft	Milligrams/cu ft	0.001
	Micrograms/liter	0.001
	ppm by volume (20° C)	0.02404/M
	ppm by weight	834.7×10^{-6}
	Pounds/cu ft	62.43×10^{-12}
Micrograms/liter	Milligrams/cu m	1.0
	Micrograms/cu m	1000.0
	ppm by volume (20°C)	24.04/M
	ppm by weight	0.8347
	Pounds/cu ft	62.43×10^{-9}
Pounds/cu ft	Milligrams/cu m	16.018×10^{6}
	Micrograms/cu m	16.018×10^{9}
	Micrograms/liter	16.018×10^{6}
	ppm by volume (20°C)	$385.1 \times 10^{6}/M$
	ppm by weight	133.7×10^{3}
ppm by volume (20°C)	Milligrams/cu m	M/24.04
	Micrograms/cu m	0.02404/M
	Micrograms/liter	M/24.04
	ppm by weight	M/28.8
	Pounds/cu ft	$M/385.1 \times 10^{6}$
ppm by weight	Milligrams/cu m	1.198
	Micrograms/cu m	1.198×10^{3}
	Micrograms/liter	1.198

- *Factor* or *divisor* of a whole number: any other whole number that exactly divides it. Thus, 2 and 5 are factors of 10.
- *Prime number*: a number that has no factors except itself and 1. Examples of prime numbers are 1, 3, 5, 7, and 11.
- *Composite number*: a number that has factors other than itself and 1. Examples of composite numbers are 4, 6, 8, 9, and 12.
- *Common factor* or *common divisor* of two or more numbers: a factor that will exactly divide each of them. If this factor is the largest factor possible, it is called the *greatest common divisor*. Thus, 3 is a common divisor of 9 and 27, and 9 is the greatest common divisor of 9 and 27.
- *Multiple* of a given number: a number that is exactly divisible by the given number. If a number is exactly divisible by two or more other numbers, it is a common

Table 1.7   Atmospheric pressure

To convert from	To	Multiply by
Atmospheres	Millimeters of mercury	760.0
	Inches of mercury	29.92
	Millibars	1013.2
Millimeters of mercury	Atmospheres	1.316×10^{-3}
	Inches of mercury	39.37×10^{-3}
	Millibars	1.333
Inches of mercury	Atmospheres	0.03333
	Millimeters of mercury	25.4005
	Millibars	33.35
Millibars	Atmospheres	0.000987
	Millimeters of mercury	0.75
	Inches of mercury	0.30

Sampling Pressures

To convert from	To	Multiply by
Millimeters of mercury (0°C)	Inches of water (60°C)	0.5358
Inches of mercury (0°C)	Inches of water (60°C)	13.609
Inches of water	Millimeters of mercury (0°C)	1.8663
	Inches of mercury (0°C)	73.48×10^{-2}

Table 1.8   Velocity

To convert from	To	Multiply by
Meters/sec	Kilometers/hr	3.6
	Feet/sec	3.281
	Miles/hr	2.237
Kilometers/hr	Meters/sec	0.2778
	Feet/sec	0.9113
	Miles/hr	0.6241
Feet/hr	Meters/sec	0.3048
	Kilometers/hr	1.0973
	Miles/hr	0.6818
Miles/hr	Meters/sec	0.4470
	Kilometers/hr	1.6093
	Feet/sec	1.4667

Table 1.9   Atmospheric particulate matter

To convert from	To	Multiply by
Milligrams/cu m	Grams/cu ft	283.2×10^{-6}
	Grams/cu m	0.001
	Micrograms/cu m	1000.0
	Monograms/cu ft	28.32
	Pounds/1000 cu ft	62.43×10^{-6}
Grams/cu ft	Milligrams/cu m	35.3145×10^{3}
	Grams/cu m	35.314
	Micrograms/cu m	35.314×10^{3}
	Micrograms/cu ft	1.0×10^{6}
	Pounds/1000 cu ft	2.2046

Table 1.10   Concentration

To convert from	To	Multiply by
Grams/cu m	Milligrams/cu m	1000.0
	Grams/cu ft	0.02832
	Micrograms/cu ft	1.0×10^{6}
	Pounds/1000 cu ft	0.06243
Micrograms/cu m	Milligrams/cu m	0.001
	Grams/cu ft	28.43×10^{-9}
	Grams/cu m	1.0×10^{-6}
	Micrograms/cu ft	0.02832
	Pounds/1000 cu ft	62.43×10^{-9}
Micrograms/cu ft	Milligrams/cu m	35.314×10^{-3}
	Grams/cu ft	1.0×10^{-6}
	Grams/cu m	35.314×10^{-6}
	Micrograms	35.314
	Pounds/1000 cu ft	2.2046×10^{-6}
Pounds/1000 cu ft	Milligrams/cu m	16.018×10^{3}
	Grams/cu ft	0.35314
	Micrograms/cu m	16.018×10^{6}
	Grams/cu m	16.018
	Micrograms/cu ft	353.14×10^{2}

multiple of them. The least (smallest) such number is called the *lowest common multiple*. Thus, 36 and 72 are common multiples of 12, 9, and 4; however, 36 is the lowest common multiple.

- *Even number*: a number exactly divisible by 2. Thus, 2, 4, 6, 8, 10, and 12 are even integers.
- *Odd number*: an integer that is not exactly divisible by 2. Thus, 1, 3, 5, 7, 9, and 11 are odd integers.

Table 1.11 Soil test conversion factors

Soil sample depth inches	Multiply ppm by
3	1
6	2
7	2.33
8	2.66
9	3
10	3.33
12	4

- *Product*: the result of multiplying two or more numbers together. Thus, 20 is the product of 4 × 5. Also, 4 and 5 are factors of 20.
- *Quotient*: the result of dividing one number by another. For example, 5 is the quotient of 20 divided by 4.
- *Dividend*: a number to be divided; a *divisor* is a number that divides. For example, in 100 ÷ 20 = 5, 100 is the dividend, 20 is the divisor, and 5 is the quotient.
- *Area*: the amount of surface an object contains or the amount of material required to cover the surface, measured in square units.
- *Base*: the bottom leg of a triangle, measured in linear units.
- *Circumference*: the distance around an object, measured in linear units. When determined for other than circles, it may be called the *perimeter* of the figure, object, or landscape.
- *Cubic units*: measurements used to express volume, cubic feet, cubic meters, and so on.
- *Depth*: the vertical distance from the bottom of a tank to the top. This is normally measured in terms of liquid depth and given in terms of sidewall depth, measured in linear units.
- *Diameter*: the distance from one edge of a circle to the opposite edge passing through the center, measured in linear units.
- *Height*: the vertical distance from the base or bottom of a unit to the top or surface.
- *Linear units*: measurements used to express distances (feet, inches, meters, yards, etc.).
- *Pi, (π)*: a number in the calculations involving circles, spheres, or cones ($\pi = 3.14$).
- *Radius*: the distance from the center of a circle to the edge, measured in linear units.
- *Sphere*: a container shaped like a ball.
- *Square units*: measurements used to express area, square feet, square meters, acres, and so forth.
- *Volume*: the capacity of the unit (how much it will hold) measured in cubic units (cubic feet, cubic meters) or in liquid volume units (gallons, liters, million gallons).
- *Width*: the distance from one side of a tank to the other, measured in linear units.

Key words for math operations include:

- *of*: to multiply
- *and*: to add

- *per*: to divide
- *less than*: to subtract

SEQUENCE OF OPERATIONS

Mathematical operations such as addition, subtraction, multiplication, and division are usually performed in a certain order or sequence. Typically, multiplication and division operations are done prior to addition and subtraction operations. In addition, mathematical operations are also generally performed from left to right using this hierarchy. The use of parentheses is also common to set apart operations that should be performed in a particular sequence.

✔ Note: We assume that the reader has a fundamental knowledge of basic arithmetic and math operations. Thus, the purpose of the following section is to provide only a brief review of the mathematical concepts and applications frequently employed by physicists.

Sequence of Operations—Rules

Rule 1: In a series of additions, the terms may be placed in any order and grouped in any way. Thus:
$$4 + 3 = 7 \text{ and } 3 + 4 = 7; (4 + 3) + (6 + 4) = 17, (6 + 3) + (4 + 4) = 17, \text{ and } 6 + (3 + 4) + 4 = 17.$$

Rule 2: In a series of subtractions, changing the order or the grouping of the terms may change the result. Thus:
$$100 - 30 = 70, \text{ but } 30 - 100 = -70; (100 - 30) - 10 = 60, \text{ but } 100 - (30 - 10) = 80.$$

Rule 3: When no grouping is given, the subtractions are performed in the order written, from left to right. Thus:
$$100 - 30 - 15 - 4 = 51; \text{ or by steps, } 100 - 30 = 70, 70 - 15 = 55, 55 - 4 = 51.$$

Rule 4: In a series of multiplications, the factors may be placed in any order and in any grouping. Thus:
$$[(2 \times 3) \times 5] \times 6 = 180 \text{ and } 5 \times [2 \times (6 \times 3)] = 180.$$

Rule 5: In a series of divisions, changing the order or the grouping may change the result. Thus:
$$100 \div 10 = 10, \text{ but } 10 \div 100 = 0.1; (100 \div 10) \div 2 = 5, \text{ but } 100 \div (10 \div 2) = 20.$$

Rule 6: Again, if no grouping is indicated, the divisions are performed in the order written, from left to right. Thus:
$$100 \div 10 \div 2 \text{ is understood to mean } (100 \div 10) \div 2.$$

Rule 7: In a series of mixed mathematical operations, the convention is as follows: whenever no grouping is given, multiplications and divisions are to be performed in the order written, then additions and subtractions in the order written.

Sequence of Operations—Examples

In a series of additions, the terms may be placed in any order and grouped in any way. Examples:

$$4 + 6 = 10 \text{ and } 6 + 4 = 10$$
$$(4 + 5) + (3 + 7) = 19, (3 + 5) + (4 + 7) = 19, \text{ and}$$
$$[7 + (5 + 4)] + 3 = 19$$

In a series of subtractions, changing the order or the grouping of the terms may change the result. Examples:

$$100 - 20 = 80, \text{ but } 20 - 100 = -80$$
$$(100 - 30) - 20 = 50, \text{ but } 100 - (30 - 20) = 90$$

When no grouping is given, the subtractions are performed in the order written—from left to right. Examples:

$$100 - 30 - 20 - 3 = 47$$

or by steps,

$$100 - 30 = 70, 70 - 20 = 50, 50 - 3 = 47$$

In a series of multiplications, the factors may be placed in any order and in any grouping. Examples:

$$[(3 \times 3) \times 5] \times 6 = 270 \text{ and } 5 \times [3 \times (6 \times 3)] = 270$$

In a series of divisions, changing the order or the grouping may change the result. Examples:

$$100 \div 10 = 10, \text{ but } 10 \div 100 = 0.1$$
$$(100 \div 10) \div 2 = 5, \text{ but } 100 \div (10 \div 2) = 20$$

If no grouping is indicated, the divisions are performed in the order written—from left to right. Example:

$$100 \div 5 \div 2 \text{ is understood to mean } (100 \div 5) \div 2$$

In a series of mixed mathematical operations, the rule of thumb is: Whenever no grouping is given, multiplications and divisions are to be performed in the order written, and then additions and subtractions in the order written.

PERCENT

The word *percent* means "by the hundred." Percentage is often designated by the symbol %. Thus, 15% means 15 percent or 15/100 or 0.15. These equivalents may be written in the reverse order: 0.15 = 15/100 = 15%. When working with percent, the following key points are important:

- Percentages are another way of expressing a part of a whole.
- As mentioned, *percent* means "by the hundred," so a percentage is the number out of 100. To determine percent, divide the quantity you wish to express as a percentage by the total quantity; then multiply by 100:

$$\text{Percent (\%)} = \frac{\text{Part}}{\text{Whole}} \tag{1.2}$$

For example, 22 percent (or 22%) means 22 out of 100, or 22/100. Dividing 22 by 100, results in the decimal 0.22:

$$22\% = \frac{22}{100} = 0.22$$

- When using percentages in calculations, the percentage must be converted to an equivalent decimal number; this is accomplished by dividing the percentage by 100. In a chemical dosing example, calcium hypochlorite (HTH) contains 65% available chlorine. What is the decimal equivalent of 65%? Because 65% means 65 per hundred divide 65 by 100: 65/100, which is 0.65.
- Decimals and fractions can be converted to percentages. The fraction is first converted to a decimal, and then the decimal is multiplied by 100 to get the percentage. For example, if a 50-foot-high water tank has 26 feet of water in it, how full is the tank in terms of the percentage of its capacity?

$$\frac{26 \text{ ft}}{50 \text{ ft}} = 0.52 \text{ (decimal equivalent)}$$

$$0.52 \times 100 = 52 \text{ (The tank is 52\% full.)}$$

Example 1.28
Problem
The plant operator removes 6,500 gal of a chemical mixture from a storage tank. The chemical mixture contains 325 gal of solids. What is the percentage of solids in the chemical mixture?

Solution

$$\text{percent} = \frac{325 \text{ gal}}{6,500 \text{ gal}} \times 100 = 5\%$$

Example 1.29
Problem
    Convert 65% to decimal percent.

Solution

$$\text{decimal percent} = \frac{\text{percent}}{100}$$

$$= \frac{65}{100} = 0.65$$

Example 1.30
Problem
A solution contains 5.8% solids. What is the concentration of solids in decimal percent?

Solution

$$\text{decimal percent} = \frac{5.8\%}{100} = 0.058$$

✔ **Key Point:** Unless otherwise noted, all calculations in the text using percent values require the percent be converted to a decimal before use.

✔ **Key Point:** To determine what quantity a percent equals, first convert the percent to a decimal, then multiply by the total quantity.

$$\text{quantity} = \text{total} \times \text{decimal percent} \tag{1.3}$$

Example 1.31
Problem
A chemical mixture drawn from the settling tank is 5% solids. If 2,800 gallons of solids are withdrawn, how many gallons of solids are removed?

Solution

$$\text{gallons} = \frac{5\%}{100} \times 2,800 \text{ gallons} = 140 \text{ gal}$$

Example 1.32
Problem
Convert 0.55 to percent.

Solution

$$0.55 = \frac{55}{100} = 0.55 = 55\%$$

In converting 0.55 to 55%, we simply moved the decimal point two places to the right.

Example 1.33
Problem
Convert 7/22 to a percent.

Solution

$$\frac{7}{22} = 0.318 \times 100 = 31.8\%$$

SIGNIFICANT DIGITS

When rounding numbers, remember the following key points:

- Numbers are rounded to reduce the number of digits to the right of the decimal point. This is done for convenience, not for accuracy.
- Rule: a number is rounded off by dropping one or more numbers from the right and adding zeroes if necessary to place the decimal point. If the last figure dropped is 5 or more, increase the last retained figure by 1. If the last digit dropped is less than 5, do not increase the last retained figure. If the digit 5 is dropped, round off the preceding digit to the nearest *even* number.

Example 1.34
Problem
Round off 10,546 to 4, 3, 2, and 1 significant figures.

Solution

10,546 = 10,550 to four significant figures
10,546 = 10,500 to three significant figures
10,546 = 11,000 to two significant figures
10,547 = 10,000 to one significant figure

Significant figures are those digits in a final calculation that have physical meaning. In determining significant figures, remember the following key points:

- The concept of significant figures is related to rounding.
- It can be used to determine where to round off.

✔ **Key Point**: No answer can be more accurate than the least accurate piece of data used to calculate the answer.

• Rule: Significant figures are those numbers that are known to be reliable. The position of the decimal point does not determine the number of significant figures.

Example 1.35
Problem
How many significant figures are in a measurement of 1.35 in?

Solution
Three significant figures: 1, 3, and 5.

Example 1.36
Problem
How many significant figures are in a measurement of 0.000135?

Solution
Again, three significant figures: 1, 3, and 5. The three zeros are used only to place the decimal point.

Example 1.37
Problem
How many significant figures are in a measurement of 103,500?

Solution
Four significant figures: 1, 0, 3, and 5. The remaining two zeros are used to place the decimal point.

POWERS AND EXPONENTS

In working with powers and exponents, important key points include:

• Powers are used to identify area (as in square feet) and volume (as in cubic feet).
• Powers can also be used to indicate that a number should be squared, cubed, etc. This later designation is the number of times a number must be multiplied times itself. For example, when several numbers are multiplied together, as $4 \times 5 \times 6 = 120$, the numbers, 4, 5, and 6 are the *factors*; 120 is the *product*.
• If all the factors are alike, as $4 \times 4 \times 4 \times 4 = 256$, the product is called a *power*. Thus, 256 is a power of 4, and 4 is the *base* of the power. A *power* is a *product* obtained by using a base a certain number of times as a factor.
• Instead of writing $4 \times 4 \times 4 \times 4$, it is more convenient to use an *exponent* to indicate that the factor 4 is used as a factor four times. This exponent, a small number placed above and to the right of the base number, indicates known many times the base is to be used as a factor. Using this system of notation, the multiplication $4 \times 4 \times 4 \times 4$ is written as 4^4. The 4 is the *exponent*, showing that 4 is to be used as a factor 4 times.

• These same consideration apply to letters (*a, b, x, y,* etc.) as well. For example:

$$z^2 = (z) (z)$$

or

$$z^4 = (z) (z) (z) (z)$$

• When a number or letter does not have an exponent, it is considered to have an exponent of 1.

The powers of 1:
$$1^0 = 1$$
$$1^1 = 1$$
$$1^2 = 1$$
$$1^3 = 1$$
$$1^4 = 1$$

The powers of 10:
$$10^0 = 1$$
$$10^1 = 10$$
$$10^2 = 100$$
$$10^3 = 1000$$
$$10^4 = 10,000$$

Example 1.38
Problem
How is the term 2^3 written in expanded form?

Solution
The power (exponent) of 3 means that the base number (2) is multiplied by itself three times.

$$2^3 = (2)(2)(2)$$

Example 1.39
Problem
How is the term $(3/8)^2$ written in expanded form?

✔ **Key Point:** When parentheses are used, the exponent refers to the entire term within the parentheses. Thus, in this example, $(3/8)^2$ means

Solution

$$(3/8)^2 = (3/8) (3/8)$$

✔ **Key Point:** When a negative exponent is used with a number or term, a number can be re-expressed using a positive exponent.

$$6^{-3} = 1/6^3$$

Another example is

$$11^{-5} = 1/11^5$$

Example 1.40
Problem
How is the term 8^{-3} written in expanded form?

$$8^{-3} = \frac{1}{8^3} = \frac{1}{(8)(8)(8)}$$

✔ **Key Point:** Any number or letter such as 3^0 or X^0 does not equal 3×1 or $X \times 1$, but simply 1.

AVERAGES (ARITHMETIC MEAN)

Whether we speak of harmonic mean, geometric mean, or arithmetic mean, each is designed to find the "center," or the "middle" of the set of numbers. They capture the intuitive notion of a central tendency that may be present in the data. In statistical analysis, an average of data is a number that indicates the middle of the distribution of data values.

An *average* is a way of representing several different measurements as a single number. Although averages can be useful by telling "about" how much or how many, they can also be misleading, as we demonstrate next.

Example 1.41
Problem
When working with averages, the *mean* (again, what we usually refer to as an average) is the total of values of a set of observations divided by the number of observations. We simply add up all of the individual measurements and divide by the total number of measurements we took. For example, the operator of a waterworks or wastewater treatment plant takes a chlorine residual measurement every day, and part of his or her operating log is shown in table 1.12. Find the mean.

Solution
Add up the seven chlorine residual readings: 0.9, 1.0, 0.9, 1.3, 1.1, 1.4, 1.2 = 7.8. Next, divide by the number of measurements, in this case seven: 7.8 ÷ 7 = 1.11. The mean chlorine residual for the week was 1.11 mg/l.

Table 1.12   Daily chlorine residual results

Day	Chlorine residual (mg/l)
Monday	0.9
Tuesday	1.0
Wednesday	0.9
Thursday	1.3
Friday	1.1
Saturday	1.4
Sunday	1.2

Example 1.42
Problem
A water system has four wells with the following capacities: 115 gpm, 100 gpm, 125 gpm, and 90 gpm. What is the mean?

Solution

$$\frac{115 \text{ gpm} + 100 \text{ gpm} + 125 \text{ gpm} + 90 \text{ gpm } 430}{4} = \frac{430}{4} = 107.5 \text{ gpm}$$

Example 1.43
Problem
A water system has four storage tanks. Three of them have a capacity of 100,000 gallons each, while the fourth has a capacity of 1 million gallons. What is the mean capacity of the storage tanks?

Solution
The mean capacity of the storage tanks is:

$$\frac{100,000 + 100,000 + 100,000 + 1,000,000}{4} = 325,000 \text{ gal}$$

Notice that no tank in Example 1.43 has a capacity anywhere close to the mean.

Example 1.44
Problem
Effluent BOD test results for the treatment plant during the month of August are shown below. What is the average effluent BOD for the month of August?

Test 1   22 mg/L
Test 2   33 mg/L
Test 3   21 mg/L
Test 4   13 mg/L

Solution

$$\text{Average} = \frac{22 \text{ mg/L} + 33 \text{ mg/L} + 21 \text{ mg/L} + 13 \text{ mg/L}}{4} = 22.3 \text{ mg/L}$$

Example 1.45
Problem

For the primary influent flow, the following composite-sampled solids concentrations were recorded for the week in the following table. What is the average sampled solids (SS)?

Monday	310 mg/L SS
Tuesday	322 mg/L SS
Wednesday	305 mg/L SS
Thursday	326 mg/L SS
Friday	313 mg/L SS
Saturday	310 mg/L SS
Sunday	320 mg/L SS
Total	2,206 mg/L SS

Solution

$$\text{Average SS} = \frac{\text{Sum of All Measurements}}{\text{Number of Measurements Used}}$$

$$= \frac{2,206 \text{ mg/L SS}}{7} = 315.1 \text{ mg/L SS}$$

SOLVING FOR THE UNKNOWN

Many calculations in physics involve the use of formulae and equations. To make these calculations, you must first know the values for all but one of the terms of the equation to be used. The obvious question is "What is an equation?" Simply, an equation is a mathematical statement that tells us that what is on one side of an equal sign (=) is equal to what is on the other side.

✔ **Key Point:** What we do to one side of the equation, we must do to the other side. This is the case, of course, because the two sides, by definition, are always equal.

An *equation* is a statement that two expressions or quantities are equal in value. The statement of equality $6x + 4 = 19$ is an equation; that is, it is algebraic short-hand for "The sum of 6 times a number plus 4 is equal to 19." It can be seen that the equation $6x + 4 = 19$ is much easier to work with than the equivalent sentence.

When thinking about equations, it is helpful to consider an equation as being similar to a balance. The equal sign tells you that two quantities are "in balance" (i.e., they are equal).

Back to equation $6x + 4 = 19$. The solution to this problem may be summarized in three steps.

Step 1: $6x + 4 = 19$
Step 2: $6x = 15$
Step 3: $x = 2.5$

✔ **Note:** Step 1 expresses the whole equation. In step 2, 4 has been subtracted from both members of the equation. In step 3, both members have been divided by 6.

✔ **Key Point:** An equation is, therefore, kept in balance (both sides of the equal sign are kept equal) by subtracting the same number from both members (sides), adding the same number to both, or dividing or multiplying by the same number.

The expression $6x + 4 = 19$ is called a *conditional equation,* because it is true only when x has a certain value. The number to be found in a conditional equation is called the *unknown number*, the *unknown quantity*, or, more briefly, the *unknown.*

✔ **Key Point:** Solving an equation is finding the value or values of the unknown that make the equation true.

Let's look a look at another equation:

$$W = F \times D \tag{1.4}$$

where

W = work
F = force
D = distance

$$\text{Work} = \text{Force (lb)} \times \text{Distance (ft or in.)}$$
$$= \text{ft-lb or in-lb}$$

Suppose we have this equation:

$$60 = (x)\,(2)$$

How can we determine the value of x? By following the axioms presented below, the solution to the unknown is quite simple.

✔ **Key Point:** It is important to point out that the following discussion includes only what the axioms are and how they work.

Axioms

1. If equal numbers are added to equal numbers, the sums are equal.
2. If equal numbers are subtracted from equal numbers, the remainders are equal.
3. If equal numbers are multiplied by equal numbers, the products are equal.
4. If equal numbers are divided by equal numbers (except zero), the quotients are equal.
5. Numbers that are equal to the same number or to equal numbers are equal to each other.
6. Like powers of equal numbers are equal.
7. Like roots of equal numbers are equal.
8. The whole of anything equals the sum of all its parts.

✔ **Note:** Axioms 2 and 4 were used to solve the equation $6x + 4 = 19$.

✔ **Key Point:** As mentioned, solving an equation is determining the value or values of the unknown number or numbers in the equation.

Example 1.46
Problem
Find the value of x if $x - 8 = 2$.

Solution
Here it can be seen by inspection that $x = 10$, but inspection does not help in solving more complicated equations. However, if we notice that to determine that $x = 10$, 8 is added to each member of the given equation, we have acquired a method or procedure that can be applied to similar but more complex problems.

Given equation:

$$x - 8 = 2$$

Add 8 to each member (axiom 1):

$$x = 2 + 8$$

Collect the terms (that is, add 2 and 8):

$$x = 10$$

Example 1.47
Problem
Solve for x, if $4x - 4 = 8$ (each side is in simplest terms)

Solution

$4x = 8 + 4$ [the term (-4) is moved to the right of the equal sign as $(+4)$]
$4x = 12$
$\dfrac{4x}{4} = \dfrac{12}{4}$ (divide both sides)
$x = 3$ (x is alone on the left and is equal to the value on the right)

Example 1.48
Problem
Solve for x, if $x + 10 = 15$.

Solution
Subtract 10 from each member (axiom 2):

$$x = 15 - 10$$

Collect the terms:

$$x = 5$$

Example 1.49
Problem
Solve for x, if $5x + 5 - 7 = 3x + 6$.

Solution
Collect the terms $(+5)$ and (-7):

$$5x - 2 = 3x + 6$$

Add 2 to both members (axiom 2):

$$5x = 3x + 8$$

Subtract $3x$ from both members (axiom 2):

$$2x = 8$$

Divide both members by 2 (axiom 4):

$$x = 4$$

After obtaining a solution to an equation, we should always check it. This is an easy
process. All we need to do is substitute the solution for the unknown quantity in the

given equation. If the two members of the equation are then identical, the number substituted is the correct answer.

Example 1.50
Problem
Solve and check $4x + 5 - 7 = 2x + 6$

Solution

$$4x + 5 - 7 = 2x + 6$$
$$4x - 2 = 2x + 6$$
$$4x = 2x + 8$$
$$2x = 8$$
$$x = 4$$

Substituting the answer $x = 4$ in the original equation,

$$4x + 5 - 7 = 2x + 6$$
$$4(4) + 5 - 7 = 2(4) + 6$$
$$16 + 5 - 7 = 8 + 6$$
$$14 = 14$$

Because the statement $14 = 14$ is true, the answer $x = 4$ must be correct.

The equations discussed to this point were expressed in *algebraic* language. It is important to learn how to set up an equation by translating a sentence into an equation (into algebraic language) and then solving the equation.

In setting up an equation properly, the following suggestions and examples should help.

1. Always read the statement of the problem carefully.
2. Select the unknown number and represent it by some letter. If more than one unknown quantity exists in the problem, try to represent those numbers in terms of the same letter—that is, in terms of one quantity.
3. Develop the equation, using the letter or letters selected, and then solve.

Example 1.51
Problem
Given: One number is 8 more than another. The larger number is 2 less than three times the smaller. What are the two numbers?

Solution
Let n represent the small number. Then $n + 8$ must represent the larger number.

$$n + 8 = 3n - 2$$
$$n = 5 \text{ (small number)}$$
$$n + 8 = 13 \text{ (large number)}$$

Example 1.52
Problem
If five times the sum of a number and 6 is increased by 3, the result is 2 less than ten times the number. Find the number.

Solution
Let *n* represent the number.

$$5(n + 6) + 3 = 10n - 2$$
$$n = 5$$

Example 1.53
Problem
If $2x + 5 = 10$, solve for *x*.

Solution

$$2x + 5 = 10$$
$$2x = 5$$
$$x = 5/2$$
$$x = 2.5$$

Example 1.54
Problem
If $.5x - 1 = -6$, find *x*.

Solution

$$.5x - 1 = -6$$
$$.5x = -5$$
$$x = -10$$

RATIO

A *ratio* is the established relationship between two numbers. For example, if someone says, "I'll give you four to one the Redskins over the Cowboys in the Super Bowl," what does that person mean?

Four to one, or 4:1, is a ratio. If someone gives you four to one, it's his or her $4 to your $1.

As another more pertinent example, an average of 3 cu ft of screenings are re-moved from each million gallons of wastewater treated; the ratio of screenings re-moved (cu ft) to treated wastewater (MG) is 3:1. Ratios are normally written using a colon (such as 2:1) or as a fraction (such as 2/1).

A *proportion* is a statement that two ratios are equal. For example, 1 is to 2 as 3 is to 6, so 1:2 = 3:6. In this case, 1 has the same relation to 2 that 3 has to 6. And what exactly is that relation? What do you think? You're right: 1 is half the size of 2, and 3 is half the size of 6. Or, alternately, 2 is twice the size of 1, and 6 is twice the size of 3.

When working with ratio and/or proportion, the following key points are impor-tant to remember.

1. One place where fractions are used in calculations is when ratios and proportions are used, such as in calculating solutions.
2. A ratio is usually stated in the form A is to B as C is to D, and we can write it as two fractions that are equal to each other:

$$\frac{A}{B} = \frac{C}{D}$$

3. Cross-multiplying solves ratio problems; that is, we multiply the left numerator (A) by the right denominator (D) and say that is equal to the left denominator (B) times the right numerator (C):

$$A \times D = B \times C$$
$$AD = BC$$

4. If one of the four items is unknown, dividing the two known items that are multi-plied together by the known item that is multiplied by the unknown solves the ratio. For example, if 2 pounds of alum are needed to treat 500 gallons of water, how many pounds of alum will we need to treat 10,000 gallons? We can state this as a ratio: 2 pounds of alum is to 500 gallons of water as x pounds of alum is to 10,000 gallons of water.

This is set up in this manner:

$$\frac{1 \text{ lb alum}}{500 \text{ gal water}} = \frac{x \text{ lb alum}}{10,000 \text{ gal water}}$$

Cross-multiplying:

$$(500)\,(x) = (1) \times (10,000)$$

Transposing:

$$x = \frac{1 \times 10,000}{500}$$

$$x = 20 \text{ lb alum}$$

For calculating proportion, for example, 5 gallons of fuel costs $5.40. How much does 15 gallons cost?

$$\frac{5 \text{ gal}}{\$5.40} = \frac{15 \text{ gal}}{\$y}$$

$$5 \times y = 15 \times 5.40 = 81$$

$$y = \frac{81}{5} = \$16.20$$

Example 1.55
Problem
If a pump will fill a tank in 20 hours at 4 gpm (gallons per minute), how long will it take a 10-gpm pump to fill the same tank?

Solution
First, analyze the problem. Here, the unknown is some number of hours. But should the answer be larger or smaller than 20 hours? If a 4-gpm pump can fill the tank in 20 hours, a larger pump (10-gpm) should be able to complete the filling in less than 20 hours. Therefore, the answer should be less than 20 hours.

Now set up the proportion:

$$\frac{x \text{ h}}{20 \text{ h}} = \frac{4 \text{ gpm}}{10}$$

$$x = \frac{(4)(20)}{10}$$

$$x = 8 \text{ h}$$

Example 1.56
Problem
Solve for *x* in the proportion problem given below.

Solution

$$\frac{36}{180} = \frac{x}{4{,}450}$$

$$\frac{(4{,}450)(36)}{180} = x$$

$$= 890$$

Example 1.57
Problem
Solve for the unknown value *x* in the problem given below.

Solution

$$\frac{3.4}{2} = \frac{6}{x}$$

$$(3.4)\,(x) = (2)\,(6)$$

$$x = \frac{(2)\,(6)}{3.4}$$

$$x = 3.53$$

Example 1.58
Problem
1 lb of chlorine is dissolved in 65 gallons of water. To maintain the same concentration, how many pounds of chlorine would have to be dissolved in 150 gallons of water?

Solution

$$\frac{1\ \text{lb}}{65\ \text{gal}} = \frac{x\ \text{lb}}{150\ \text{gal}}$$

$$(65)\,(x) = (1)\,(150)$$

$$x = \frac{(1)\,(150)}{65}$$

$$= 2.3\ \text{lb}$$

Example 1.59
Problem
It takes 5 workers 50 hours to complete a job. At the same rate, how many hours would it take 8 workers to complete the job?

Solution

$$\frac{5\ \text{workers}}{8\ \text{workers}} = \frac{x\ \text{hours}}{50\ \text{hours}}$$

$$x = \frac{(5)\,(50)}{8}$$

$$x = 31.3\ \text{hours}$$

DIMENSIONAL ANALYSIS

Dimensional analysis is a problem-solving method that uses the fact that 1 can multiply any number or expression, without changing its value. It is a useful technique used to check if a problem is set up correctly. In using dimensional analysis to check a math setup, we work with the dimensions (units of measure) only—not with numbers.

An example of dimensional analysis common to everyday life is the unit pricing found in many hardware stores. A shopper can purchase a 1-lb box of nails for $0.98 in one store, whereas a warehouse store sells a 5-lb bag of the same nails for $3.50. The shopper will analyze this problem almost without thinking about it. The solution calls for reducing the problem to the price per pound. The pound is selected as the unit common to both stores. Knowing the unit price, which is expressed in dollars per pound ($/lb), is implicit in the solution to this problem.

To use the dimensional analysis method, we must know how to perform three basic operations:

✔ **Note:** Unit factors may be made from any two terms that describe the same or equivalent "amounts" of what we are interested in. For example, we know that 1 inch = 2.54 centimeters.

1. *Basic Operation:* To complete a division of units, always ensure that all units are written in the same format; it is best to express a horizontal fraction (such as gal/ft²) as a vertical fraction. Horizontal to vertical:

$$\text{gal/ft}^3 \text{ to } \frac{\text{gal}}{\text{ft}^3}$$

$$\text{psi to } \frac{\text{lb}}{\text{in.}^2}$$

2. *Basic Operation:* We must know how to divide by a fraction. For example,

$$\frac{\dfrac{\text{lb}}{\text{d}}}{\dfrac{\text{min}}{\text{d}}} \text{ becomes } \frac{\text{lb}}{\text{d}} \times \frac{\text{d}}{\text{min}}$$

In the preceding example, notice that the terms in the denominator were inverted before the fractions were multiplied. This is a standard rule that must be followed when dividing fractions. An example is

$$\frac{\dfrac{\text{mm}^2}{\text{mm}^2}}{\text{m}^2} \text{ becomes } \text{mm}^2 \times \frac{\text{m}^2}{\text{mm}^2}$$

3. *Basic Operation*: We must know how to cancel or divide terms in the numerator and denominator of a fraction. After fractions have been rewritten in the vertical form and division by the fraction has been re-expressed as multiplication as shown above, then the terms can be canceled (or divided) out.

📌 **Key Point:** For every term that is canceled in the numerator of a fraction, a similar term must be canceled in the denominator and vice versa, as shown below:

$$\frac{Kg}{d} \times \frac{d}{min} = \frac{kg}{min}$$

$$mm^2 \times \frac{m^2}{mm^2} = m^2$$

$$\frac{gal}{min} \times \frac{ft^3}{gal} = \frac{ft^3}{min}$$

How are units that include exponents calculated?

When written with exponents, such as ft^3, a unit can be left as is or put in expanded form, e.g., (ft)(ft)(ft), depending on other units in the calculation. The point is that it is important to ensure that square and cubic terms are expressed uniformly, as sq ft and cu ft, or as ft^2 and ft^3. For dimensional analysis, the latter system is preferred.

For example, if we wish to convert 1,400 ft^3 volume to gallons, we will use 7.48 gal/ft^3 in the conversions. The question becomes do we multiply or divide by 7.48?

In the above instance, it is possible to use dimensional analysis to answer this question—that is, are we to multiply or divide by 7.48? In order to determine if the math setup is correct, only the dimensions are used.

First, try dividing the dimensions:

$$\frac{ft^3}{gal/ft^3} = \frac{\dfrac{ft^3}{gal}}{ft^3}$$

Then, the numerator and denominator are multiplied to get

$$= \frac{ft^6}{gal}$$

Thus, by dimensional analysis we determine that if we divide the two dimensions (ft^3 and gal/ft^3), the units of the answer are ft^6/gal, not gal. It is clear that division is not the right way to go in making this conversion.

What would have happened if we had multiplied the dimensions instead of dividing?

$$(ft^3) \ (gal/ft^3) \ \left(\frac{gal}{ft^3}\right)$$

Then, multiply the numerator and denominator to obtain

$$= \frac{(ft^3) \ (gal)}{ft^3}$$

and cancel common terms to obtain

$$= \frac{(ft^3) \ (gal)}{ft^3}$$

Obviously, by multiplying the two dimensions (ft^3 and gal/ft^3), the answer will be in gallons, which is what we want. Thus, because the math setup is correct, we would then multiply the numbers to obtain the number of gallons.

$$(1{,}400 \ ft^3) \ (7.48 \ gal/ft^3) \ = \ 10{,}472 \ gal$$

Now let's try another problem with exponents. We wish to obtain an answer in square feet. If we are given the two terms—70 ft^3/s and 4.5 ft/s—is the following math setup correct?

$$(70 \ ft^3/sec) \ (4.5 \ ft/sec)$$

First, only the dimensions are used to determine if the math setup is correct. Multiplying the two dimensions yields:

$$(ft^3/s) \ (ft/s) \ = \ \left(\frac{ft^3}{s}\right) \left(\frac{ft^3}{s}\right)$$

Then, the terms in the numerators and denominators of the fraction are multiplied:

$$= \frac{(ft^3) \ (ft)}{(s) \ (s)}$$

$$= \frac{ft^4}{s^2}$$

Obviously, the math setup is incorrect because the dimensions of the answer are not square feet. Therefore, if we multiply the numbers as shown above, the answer will be wrong.

Let's try division of the two dimensions instead.

$$\text{ft}^3/\text{sec} = \frac{\dfrac{\text{ft}^3}{\text{sec}}}{\dfrac{\text{ft}}{\text{sec}}}$$

Invert the denominator and multiply to get

$$= \left(\frac{\text{ft}^3}{(\text{sec})}\right)\left(\frac{\text{sec}}{(\text{ft})}\right)$$

$$= \frac{(\text{ft})(\text{ft})(\text{ft})(\text{sec})}{(\text{sec})(\text{ft})}$$

$$= \frac{(\text{ft})\ (\text{ft})\ (\text{ft})\ (\text{sec})}{(\text{sec})\ (\text{ft})}$$

$$= \text{ft}^2$$

Because the dimensions of the answer are square feet, this math setup is correct. Therefore, by dividing the numbers as was done with units, the answer will also be correct.

$$\frac{70 \text{ ft}^3 \text{ /s}}{4.5 \text{ ft/s}} = 15.56 \text{ ft}^2$$

Example 1.60
Problem
We are given two terms 5 m/s and 7 m², and the answer to be obtained is in cubic meters per second (m^3/s). Is multiplying the two terms the correct math setup?

Solution

$$(\text{m/sec})\ (\text{m}^2) = \frac{\text{m}^2}{\text{sec}} \times \text{m}^2$$

Multiply the numerators and denominator of the fraction:

$$= \frac{(\text{m})\ (\text{m}^2)}{\text{sec}}$$

$$= \frac{\text{m}^3}{\text{sec}}$$

Because the dimensions of the answer are cubic metes per second (m³/s), the math setup is correct. Therefore, multiply the numbers to get the correct answer.

$$5 \text{ (m/sec) (7 m}^2) = 35 \text{ m}^3\text{/sec}$$

Example 1.61
Problem
Solve the following problem: The flow rate in a water line is 2.3 ft³/sec. What is the flow rate expressed as gallons per minute?

Solution
Set up the math problem:

$$(1.3 \text{ ft}^3 \text{/sec) (7.48 gal/ft}^3) (60 \text{ sec/min})$$

Then, use dimensional analysis to check the math setup:

$$(\text{ft}^3\text{/sec) (gal/ft}^3) (\text{sec/min}) = \left(\frac{\text{ft}^3}{\text{sec}}\right) \left(\frac{\text{gal}}{\text{ft}^3}\right) \left(\frac{\text{sec}}{\text{min}}\right)$$

$$= \frac{\text{ft}^3}{\text{sec}} \frac{\text{gal}}{\text{ft}^3} \frac{\text{sec}}{\text{min}}$$

$$= \frac{\text{gal}}{\text{min}}$$

The math setup is correct as shown above. Therefore, this problem can be multiplied out to got the answer in correct units.

$$(1.3 \text{ ft}^3\text{/sec) (7.48 gal/ft}^3) (60 \text{ sec/min}) = 1{,}032.24 \text{ gal/min}$$

MASS

The terms *weight* and *mass* are often confused. The weight of an object is dependent on its mass and the gravitational force pulling on it. This is why our weight on the moon would be less; it exerts less gravitational pull on our bodies. Weight is a measure of how heavy an object is, measured in units of force. However, in regard to mass (a fundamental concept in physics), on the moon it does not change or vary. This is the case because mass is a measure of the number of atoms in an object. Simply, unless we add to or subtract atoms from our bodies (by eating too much and/or dieting, for example), it does not matter what we are doing or where we are—our mass remains the same.

The fundamental (SI) unit of mass is the kilogram (kg). Many other units of mass are also employed, such as:

- gram: 1 g = 0.001 kg
- ton: 1 ton = 1,000 kg
- eV/c²

SPECIFIC GRAVITY AND DENSITY

Specific gravity is the ratio of the *density* (ratio of a body's mass to its volume; in layman's terms, density is a measurement of how much mass is packed into a given volume—Density = Mass/Volume) of a substance to that of a standard material under standard conditions of temperature and pressure. The specific gravity of water is 1.0. Any substance with a density greater than that of water will have a specific gravity greater than 1.0, and any substance with a density less than that of water will have a specific gravity less than 1.0. Specific gravity can be used to calculate the weight of a gallon of liquid chemical.

$$\text{Chemical, lb/gal} = \text{Water, lb/gal} \times \text{Specific Gravity (chemical)} \qquad (1.5)$$

Example 1.62
Problem
The label states that the ferric chloride solution has a specific gravity of 1.58. What is the weight of 1 gallon of ferric chloride solution?

Solution

$$\text{Ferric chloride} = 8.34 \text{ lb/gal} \times 1.58 = 13.2 \text{ lb/gal}$$

Example 1.63
Problem
If we say that the density of gasoline is 43 lb/cu ft, what is the specific gravity of gasoline?

Solution
The specific gravity of gasoline is the comparison—or ratio—of the density of gasoline to that of water:

$$\text{Specific Gravity} = \frac{\text{Density of Gasoline}}{\text{Density of Water}} = \frac{43 \text{ lb/cu ft (density of gasoline)}}{62.4 \text{ lb/cu ft (density of water)}} = 0.69$$

✔ **Key Point:** Because gasoline's specific gravity is less than 1.0 (lower than water's specific gravity), it will float in water. If gasoline's specific gravity were greater than water's specific gravity, it would sink.

✔ **Key Point:** The density of water is 1,000 g per 1,000 cubic centimeters (cm^3)—or, more simply, 1,000/1,000 = 1 gram per cubic centimeter (g/cm^3)—at a temperature of 4°C.

FLOW

Flow is expressed in many different terms in the English system of measurements. The most commonly used flow terms are as follows:

- gpm—gallons per minute
- cfs—cubic feet per second
- gpd—gallons per day
- MGD—million gallons per day

In converting flow rates, the most common flow conversions are 1 cfs = 448 gpm and 1 gpm = 1,440 gpd.

To convert gallons per day to MGD, divide the gpd by 1,000,000. For instance, convert 150,000 gallons to MGD:

$$\frac{150,000 \text{ gpd}}{1,000,000} = 0.150 \text{ MGD}$$

In some instances, flow is given in MGD but is needed in gpm. To make the conversion (MGD to gpm), two steps are required.

Step 1: Convert the gpd by multiplying by 1,000,000.
Step 2: Convert to gpm by dividing by the number of minutes in a day (1,440 min/day).

Example 1.64
Problem
Convert 0.135 MGD to gpm.

Solution
First convert the flow in MGD to gpd.

$$0.135 \text{ MGD} \times 1,000,000 = 135,000 \text{ gpd}$$

Now convert to gpm by dividing by the number of minutes in a day (24 hrs per day × 60 min per hour) = 1,440 min/day.

$$\frac{135,000 \text{ gpd}}{1,440 \text{ min/day}} = 93.8 \text{ or } 94 \text{ gpm}$$

In determining flow through a pipeline, channel, or stream, we use the following equation:

$$Q = VA \tag{1.6}$$

where

Q = cubic feet per second (cfs)
V = velocity in feet per second (ft/sec)
A = area in square feet (ft²)

Example 1.65
Problem
Find the flow in cfs in an 8-inch line, if the velocity is 3 feet per second.

Solution

Step 1: Determine the cross-sectional area of the line in square feet. Start by converting the diameter of the pipe to inches.

Step 2: The diameter is 8 inches; therefore, the radius is 4 inches. 4 inches is 4/12 of a foot or 0.33 feet.

Step 3: Find the area in square feet.

$$A = \pi r^2$$
$$A = \pi (0.33 \text{ ft})^2$$
$$A = \pi \times 0.109 \text{ ft}^2$$
$$A = 0.342 \text{ ft}^2$$

Step 4: Q = VA

$$Q = 3 \text{ ft/sec} \times 0.342 \text{ ft}^2$$
$$Q = 1.03 \text{ cfs}$$

Example 1.66
Problem
Find the flow in gpm when the total flow for the day is 75,000 gpd.

Solution

$$\frac{75,000 \text{ gpd}}{1,440 \text{ min/day}} = 52 \text{ gpm}$$

Example 1.67
Problem
Find the flow in gpm when the flow is 0.45 cfs.

Solution

$$0.45 \frac{\text{cfs}}{1} \times \frac{448 \text{ gpm}}{1 \text{cfs}} = 202 \text{ gpm}$$

HORSEPOWER AND WORK

Horsepower is a common expression for power. One horsepower is equal to 33,000 foot-pounds of work per minute. This value is determined, for example, for selecting a pump or combination of pumps to ensure an adequate pumping capacity. Pumping capacity depends upon the flow rate desired and the feet of head against which the pump must pump (a.k.a., effective height).

Calculations of horsepower are made in conjunction with many industrial operations. The basic concept from which the horsepower calculation is derived is the concept of work. *Work* involves the operation of a force (lb) over a specific distance (ft). The *amount of work* accomplished is measured in foot-pounds:

$$(ft) \ (lb) \ = \ ft\text{-}lb \tag{1.7}$$

The *rate of doing work (power)* involves a time factor. Originally, the rate of doing work or power compared the power of a horse to that of a steam engine. The rate at which a horse could work was determined to be about 550 ft-lb/sec (or expressed as 33,000 ft-lb/min). This rate has become the definition of the standard unit called horsepower.

$$\text{Horsepower, hp} \ = \ \frac{\text{Power, ft-lb/min}}{33,000 \ \text{ft-lb/min/HP}} \tag{1.8}$$

When the major use of horsepower calculation is to determine the proper pumping station operation, the horsepower calculation can be modified as shown below.

Water Horsepower (Whp)

The amount of power required to move a given volume of water is specified as total head and is known as *water horsepower (Whp)*.

$$\text{Whp} \ = \ \frac{\text{Pump Rate, gpm} \times \text{Total Head, ft} \times 8.34 \ \text{lb/gal}}{33,000 \ \text{ft-lb/min/HP}} \tag{1.9}$$

Example 1.68
Problem
A pump must deliver 1,210 gpm to total head of 130 feet. What is the required water horsepower?

Solution

$$\text{Whp} \ = \ \frac{1,210 \ \text{gpm} \times 130 \ \text{ft} \times 8.34 \ \text{lb/gal}}{33,000 \ \text{ft-lb/min/HP}}$$

Brake Horsepower (bhp)

Brake horsepower (bhp) refers to the horsepower supplied to the pump from the motor. As power moves through the pump, additional horsepower is lost from slippage and friction of the shaft and other factors; thus, pump efficiencies range from about 50% to 85% and pump efficiency must be taken into account.

$$Bhp = \frac{Whp}{Pump \text{ \% Efficiency}} \qquad (1.10)$$

Example 1.69
Problem
Under the specified conditions, the pump efficiency is 73%. If the required water horsepower is 40 hp, what is the required brake horsepower?

Solution

$$Bhp = \frac{40 \; Whp}{0.73} = 55 \; Bhp$$

Motor Horsepower

Motor horsepower (Mhp) is the horsepower the motor must generate to produce the desired brake and water horsepower.

$$Mhp = \frac{Brake \; Horsepower, \; Bhp}{Motor \text{ \% Efficiency}} \qquad (1.11)$$

Example 1.70
Problem
The motor is 93% efficient. What is the required motor horsepower when the required brake horsepower is 49.0 Bhp?

Solution

$$Mhp = \frac{49 \; Bhp}{0.93} = 53 \; Mhp$$

AREA, VOLUME, AND DENSITY

To aid in performing these calculations, the following definitions are provided.

- *Area*: the area of an object, measured in square units.
- *Base*: the term used to identity the bottom leg of a triangle, measured in linear units.
- *Circumference*: the distance around an object, measured in linear units. When determined for other than circles, it may be called the perimeter of the figure, object, or landscape.

- *Cubic units*: measurements used to express volume, cubic feet, cubic meters, and so on.
- *Depth*: the vertical distance from the bottom of a tank to the top. Normally measured in terms of liquid depth and given in terms of side wall depth (SWD), measured in linear units.
- *Diameter*: the distance from one edge of a circle to the opposite edge passing through the center, measured in linear units.
- *Height*: the vertical distance from one end of an object to the other, measured in linear units.
- *Length*: the distance from one end of an object to the other, measured in linear units.
- *Linear units*: measurements used to express distances: feet, inches, meters, yards, and so on.
- *Pi, π*: a number in the calculations involving circles, spheres, or cones ($\pi = 3.14$).
- *Radius*: the distance from the center of a circle to the edge, measured in linear units.
- *Sphere*: a container shaped like a ball.
- *Square units*: measurements used to express area, square feet, square meters, acres, and so on.
- *Volume*: the capacity of the unit; how much it will hold, measured in cubic units (cubic feet, cubic meters) or in liquid volume units (gallons, liters, million gallons).
- *Width*: the distance from one side of a tank to the other, measured in linear units.

On occasion, it may be necessary to determine the distance around grounds or landscapes. In order to measure the distance around property, buildings, and basin-like structures, it is necessary to determine either perimeter or circumference. The *perimeter* is the distance around an object—a border or outer boundary. *Circumference* is the distance around a circle or circular object, such as a clarifier. Distance is linear measurement, which defines the distance (or length) along a line. Standard units of measurement like inches, feet, yards, and miles and metric units like centimeters, meters, and kilometers are used.

PERIMETER

The perimeter of a rectangle (a four-sided figure with four right angles) is obtained by adding the lengths of the four sides (see figure 1.1).

$$\text{Perimeter} = L_1 + L_2 + L_3 + L_4 \tag{1.12}$$

Example 1.71
Problem
Find the perimeter of the rectangle shown in figure 1.2.

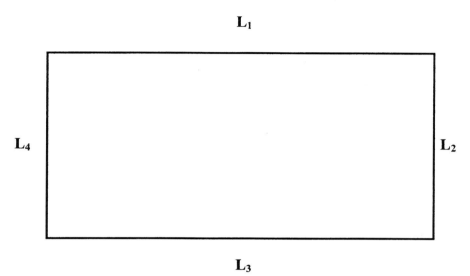

Figure 1.1. Perimeter.

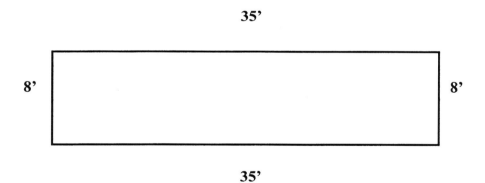

Figure 1.2. (For example 1.71.)

Solution

$$P = 35 \text{ ft} + 8 \text{ ft} + 35 \text{ ft} + 8 \text{ ft}$$
$$P = 86 \text{ ft}$$

Example 1.72
Problem
What is the perimeter of a rectangular field if its length is 100 feet and its width is 50 feet?

Solution

$$\begin{aligned}
\text{Perimeter} &= 2 \times \text{length} + 2 \times \text{width} \\
&= 2 \times 100 \text{ ft} + 2 \times 50 \text{ ft} \\
&= 200 \text{ ft} + 100 \text{ ft} \\
&= 300 \text{ ft}
\end{aligned}$$

Example 1.73
Problem
What is the perimeter of a square whose side is 8 inches?

Solution

$$\begin{aligned}
\text{Perimeter} &= 2 \times \text{length} + 2 \times \text{width} \\
&= 2 \times 8 \text{ in.} + 2 \times 8 \text{ in.} \\
&= 16 \text{ in.} + 16 \text{ in.} \\
&= 32 \text{ in.}
\end{aligned}$$

CIRCUMFERENCE

The circumference is the length of the outer border of a circle. The circumference is found by multiplying pi (π) times the diameter (D) (the diameter is a straight line passing through the center of a circle—the distance across the circle; see figure 1.3).

$$C = \pi D \tag{1.13}$$

where

C = circumference
π = Greek letter pi = 3.1416
D = diameter

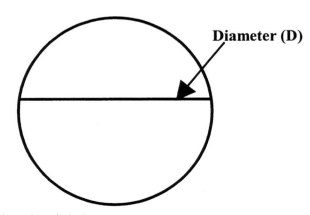

Figure 1.3. Diameter of circle.

Use this calculation if, for example, the circumference of a circular tank must be determined.

Example 1.74
Problem
Find the circumference of a circle that has a diameter of 25 feet ($\pi = 3.14$).

Solution

$$C = \pi \times 25 \text{ ft}$$
$$C = 3.14 \times 25 \text{ ft}$$
$$C = 78.5 \text{ ft}$$

Example 1.75
Problem
A circular chemical holding tank has a diameter of 18 m. What is the circumference of this tank?

Solution

$$C = \pi \, 18 \text{ m}$$
$$C = (3.14)(18 \text{ m})$$
$$C = 56.52 \text{ m}$$

Example 1.76
Problem
An influent pipe inlet opening has a diameter of 6 ft. What is the circumference of the inlet opening in inches?

Solution

$$C = \pi \times 6 \text{ ft}$$
$$C = 3.14 \times 6 \text{ ft}$$
$$C = 18.84 \text{ ft}$$

AREA

For area measurements, three basic shapes are particularly important: circles, rectangles, and triangles. *Area* is the amount of surface an object contains or the amount of material it takes to cover the surface. The area on top of a chemical tank is called the *surface area*. The area of the end of a ventilation duct is called the *cross-sectional area* (the area at right angles to the length of ducting). Area is usually expressed in square units, such as square inches (in.^{2}) or square feet (ft^2). Land may also be expressed in terms of square miles (sections) or acres (43,560 ft^2) or in the metric system as *hectares*.

A *rectangle* is a two-dimensional box. The area of a rectangle is found by multiplying the length (L) times width (W); see figure 1.4.

$$\text{Area} = L \times W \tag{1.14}$$

Example 1.77
Problem
Find the area of the rectangle shown in figure 1.5.

Solution

$$
\begin{aligned}
\text{Area} &= L \times W \\
&= 14 \text{ ft} \times 6 \text{ ft} \\
&= 84 \text{ ft}^2
\end{aligned}
$$

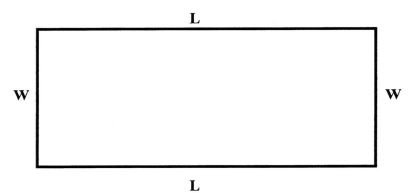

Figure 1.4.   Rectangle.

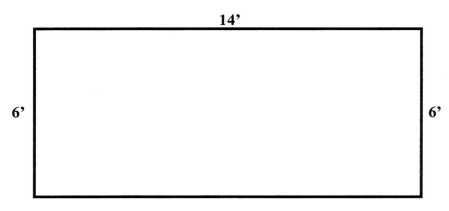

Figure 1.5.   (For example 1.77.)

Area of a Circle

To find the area of a circle, we need to introduce one new term, the *radius,* which is represented by *r*. In figure 1.6, we have a circle with a radius of 5 inches.

The radius is any straight line that radiates from the center of the circle to some point on the circumference. By definition, all radii (*radii* is the plural of radius) of the same circle are equal. The surface area of a circle is determined by multiplying _ times the radius squared.

$$\text{Area of Circle} = \pi r^2 \qquad\qquad (1.15)$$

where

A = area
π = pi (3.14)
r = radius of circle (the radius is one-half the diameter)

Example 1.78
Problem
What is the area of the circle shown in figure 1.6?

Solution

$$
\begin{aligned}
\text{Area of circle} &= \pi r^2 \\
&= \pi 6^2 \\
&= 3.14 \times 36 \\
&= 113 \text{ ft}^2
\end{aligned}
$$

Area of a Circular or Cylindrical Tank

If we were assigned to paint a water storage tank, we must know the surface area of the walls of the tank—we need to know how much paint is required. To determine

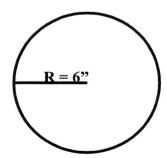

Figure 1.6.   (For example 1.78.)

the tank's surface area, we need to visualize the cylindrical walls as a rectangle wrapped around a circular base. The area of a rectangle is found by multiplying the length by the width; in this case, the width of the rectangle is the height of the wall, and the length of the rectangle is the distance around the circle, the circumference. Thus, the area of the side walls of the circular tank is found by multiplying the circumference of the base (C = π × D) times the height of the wall (H):

$$
\begin{aligned}
A &= \pi \times D \times H \\
A &= \pi \times 20 \text{ ft} \times 25 \text{ ft} \\
A &= 3.14 \times 20 \text{ ft} \times 25 \text{ ft} \\
A &= 1570.8 \text{ ft}^2
\end{aligned}
\tag{1.16}
$$

To determine the amount of paint needed, remember to add the surface area of the top of the tank, which is 314 ft^2. Thus, the amount of paint needed must cover 1570.8 ft^2 + 314 ft^2 = 1884.8 or 1,885 ft^2. If the tank floor should be painted, add another 314 ft^2.

VOLUME

The amount of space occupied by or contained in an object, *volume* (see figure 1.7), is expressed in cubic units, such as cubic inches (in.^{3}), cubic feet (ft^3), acre feet (1 acre foot = 43,560 ft^3), and so on.

Volume of a rectangular object

The volume of a rectangular object is obtained by multiplying the length times the width times the depth or height.

$$
V = L \times W \times H
\tag{1.17}
$$

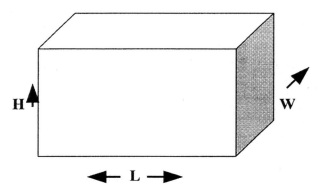

Figure 1.7. Volume of rectangle.

where

L = length
W = width
D or H = depth or height

Example 1.79
Problem
A unit rectangular process basin has a length of 15 ft, width of 7 ft, and depth of 9 ft. What is the volume of the basin?

Solution

$$V = L \times W \times D$$
$$= 15 \text{ ft} \times 7 \text{ ft} \times 9 \text{ ft}$$
$$= 945 \text{ ft}^3$$

For determining representative surface areas of rectangles, triangles, circles, or a combination of these, the practical volume formulas used are given in table 1.13.

Example 1.80
Problem
Find the volume of a 3-in. round pipe that is 300 feet long.

Step 1: Change the diameter of the pipe from inches to feet by dividing by 12.

$$D = 3 \text{ in.} \div 12 \text{ in.} = 0.25 \text{ ft}$$

Step 2: Find the radius by dividing the diameter by 2.

$$R = 0.25 \text{ ft} \div 2 = 0.125 \text{ ft}$$

Step 3: Find the volume.

Table 1.13   Volume formulas

Sphere volume	=	$(\pi/6)$ (diameter)^3
Cone volume	=	1/3 (volume of a cylinder)
Rectangular tank volume	=	(area of rectangle) (D or H)
	=	(LW) (D or H)
Cylinder volume	=	(area of cylinder) (D or H)
	=	π^2 (D or H)
Triangle volume	=	(area of triangle) (D or H)
	=	(bh/2) (D or H)

Solution

$$V = L \times \pi r^2$$
$$V = 300 \text{ ft} \times 3.14 \times 0.0156 \text{ ft}$$
$$V = 14.70 \text{ ft}^2$$

Example 1.81
Problem
Find the volume of a smokestack that is 24 in. in diameter (entire length) and 96 in. tall.

Step 1: Find the radius of the stack. The radius is one-half the diameter.

$$24 \text{ in.} \div 2 = 12 \text{ in.}$$

Step 2: Find the volume.

Solution

$$V = H \times \pi r^2$$
$$V = 96 \text{ in.} \times \pi (12 \text{ in.})^2$$
$$V = 96 \text{ in.} \times 3.41 \, (144 \text{ in.}^2)$$
$$V = 43,407 \text{ ft}^3$$

Volume of a Cone

$$\text{Volume of cone} = \frac{\pi}{12} \times \text{Diameter} \times \text{Diameter} \times \text{Height} \qquad (1.18)$$

$$\frac{\pi}{12} = \frac{3.14}{12} = 0.262$$

✔ **Key Point:** The diameter used in the formula is the diameter of the base of the cone.

Example 1.82
Problem
The bottom section of a circular settling tank has the shape of a cone. How many cubic feet of water are contained in this section of the tank if the tank has a diameter of 120 feet and the cone portion of the unit has a depth of 6 feet?

Solution

$$\text{Volume, ft}^3 = 0.262 \times 120 \text{ ft} \times 120 \text{ ft} \times 6 \text{ ft} = 22{,}637 \text{ ft}^3$$

Volume of a Sphere

$$\text{Volume} = \frac{3.14}{6} \times \text{Diameter} \times \text{Diameter} \times \text{Diameter} \qquad (1.19)$$

$$\frac{\pi}{6} = \frac{31.4}{6} = 0.52$$

Example 1.83
Problem

What is the volume of cubic feet of a gas storage container that is spherical and has a diameter of 60 feet?

Solution

$$\text{Volume, ft}^3 = 0.524 \times 60 \text{ ft} \times 60 \text{ ft} \times 60 \text{ ft} = 113{,}184 \text{ ft}^3$$

Volume of a Circular or Cylindrical Tank

A circular tank consists of a circular floor surface with a cylinder rising above it (see figure 1.8). The volume of a circular tank is calculated by multiplying the surface area times the height of the tank walls.

Example 1.84
Problem

If a tank is 20 feet in diameter and 25 feet deep, how many gallons of water will it hold?

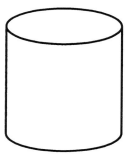

Figure 1.8. Volume of circular tank.

✔ **Hint:** In this type of problem, calculate the surface area first, multiply by the height, and then convert to gallons.

Solution

$$r = D \div 2 = 20 \text{ ft} \div 2 = 10 \text{ ft}$$

$$A = \pi \times r^2$$
$$A = \pi \times 10 \text{ ft} \times 10 \text{ ft}$$
$$A = 314 \text{ ft}^2$$

$$V = A \times H$$
$$V = 314 \text{ ft}^2 \times 25 \text{ ft}$$
$$V = 7{,}850 \text{ ft}^3 \times 7.5 \text{ gal/ft}^3 = 58{,}875 \text{ gal}$$

FORCE, PRESSURE, AND HEAD

Before we study calculations involving force, pressure, and head, we must first define these terms.

- *Force*: the push exerted by water on any confined surface. Force can be expressed in pounds, tons, grams, or kilograms.
- *Pressure*: the force per unit area. The most common way of expressing pressure is in pounds per square inch (psi).
- *Head*: the vertical distance or height of water above a reference point. Head is usually expressed in feet. In the case of water, head and pressure are related.

   A cubical container measuring one foot on each side can hold one cubic foot of water. A basic fact of science states that one cubic foot of water weighs 62.4 pounds and contains 7.48 gallons. The force acting on the bottom of the container would be 62.4 pounds per square foot. The area of the bottom in square inches is:

$$1 \text{ ft}^2 = 12 \text{ in.} \times 12 \text{ in.} = 144 \text{ in}^2$$

Therefore the pressure in pounds per square inch (psi) is:

$$\frac{62.4 \text{ lb/ft}^2}{1 \text{ ft}^2} = \frac{62.4 \text{ lb/ft}^2}{144 \text{ in}^2/\text{ft}^2} = 0.433 \text{ lb/in}^2$$

If we use the bottom of the container as our reference point, the head would be one foot. From this we can see that one foot of head is equal to 0.433 psi—an important parameter to remember.

✔ **Important Point:** *Force* acts in a particular direction. Water in a tank exerts force down on the bottom and out of the sides. *Pressure*, however, acts in all directions.

A marble at a water depth of one foot would have 0.433 psi of pressure acting inward on all sides.

Using the preceding information, we can develop equations 1.20 and 1.21 for calculating pressure and head.

$$\text{Pressure (psi)} = 0.433 \times \text{Head (ft)} \tag{1.20}$$

$$\text{Head (ft)} = 2.31 \times \text{Pressure (psi)} \tag{1.21}$$

As mentioned, *head* is the vertical distance the water must be lifted from the supply tank or unit process to the discharge. The total head includes the vertical distance the liquid must be lifted (static head), the loss to friction (friction head), and the energy required to maintain the desired velocity (velocity head).

$$\text{Total Head} = \text{Static Head} + \text{Friction Head} + \text{Velocity Head} \tag{1.22}$$

Static head is the actual vertical distance the liquid must be lifted.

$$\text{Static Head} = \text{Discharge Elevation} - \text{Supply Elevation} \tag{1.23}$$

Example 1.85
Problem
The supply tank is located at elevation 108 ft. The discharge point is at elevation 205 ft. What is the static head in feet?

Solution

$$\text{Static Head, ft} = 205 \text{ ft} - 108 \text{ ft} = 97 \text{ ft}$$

Friction head is the equivalent distance of the energy that must be supplied to overcome friction. Engineering references include tables showing the equivalent vertical distance for various sizes and types of pipes, fittings, and valves. The total friction head is the sum of the equivalent vertical distances for each component.

$$\text{Friction Head, ft} = \text{Energy Losses Due to Friction} \tag{1.24}$$

Velocity head is the equivalent distance of the energy consumed in achieving and maintaining the desired velocity in the system.

$$\text{Velocity Head, ft} = \text{Energy Losses to Maintain Velocity} \tag{1.25}$$

The pressure exerted by water is directly proportional to its depth or head in the pipe, tank, or channel. If the pressure is known, the equivalent head can be calculated.

$$\text{Head, ft} = \text{Pressure, psi} \times 2.31 \text{ ft/psi} \qquad (1.26)$$

Example 1.86
Problem
The pressure gauge on the discharge line form the influent pump reads 75.3 psi. What is the equivalent head in feet?

Solution

$$\text{Head, ft} = 75.3 \times 2.31 \text{ ft/psi} = 173.9 \text{ ft}$$

If the head is known, the equivalent pressure can be calculated by:

$$\text{Pressure, psi} = \frac{\text{Head, ft}}{2.31 \text{ ft/psi}} \qquad (1.27)$$

Example 1.87
Problem
The tank is 15 feet deep. What is the pressure in psi at the bottom of the tank when it is filled with wastewater?

Solution

$$\text{Pressure, psi} = \frac{15 \text{ ft}}{2.31 \text{ ft/psi}}$$

$$= 6.49 \text{ psi}$$

Before we look at a few example problems dealing with force, pressure, and head, it is important to list the key points related to force, pressure, and head.

1. By definition, water weighs 62.4 pounds per cubic foot.
2. The surface of any one side of the cube contains 144 square inches (12 in. $\times$ 12 in. $= 144$ in²). Therefore, the cube contains 144 columns of water 1 foot tall and 1 inch square.
3. The weight of each of these pieces can be determined by dividing the weight of the water in the cube by the number of square inches.

$$\text{Weight} = \frac{62.4 \text{ lbs}}{144 \text{ in}^2} = 0.433 \text{ lbs/in}^2 \text{ or } 0.433 \text{ psi}$$

4. Because this is the weight of one column of water one foot tall, the true expression would be 0.433 pounds per square inch per foot of head, or 0.433 psi/ft.

✔ **Key Point:** 1 foot of head = 0.433 psi.

In addition to remembering the important parameter 1 foot of head = 0.433 psi, it is important to understand the relationship between pressure and feet of head—in other words, how many feet of head 1 psi represents. This is determined by dividing 1 by 0.433.

$$\text{Feet of head} = \frac{1 \text{ ft}}{0.433 \text{ psi}} = 2.31 \text{ ft/psi}$$

If a pressure gauge were reading 12 psi, the height of the water necessary to represent this pressure would be 12 psi × 2.31 ft/psi = 27.7 feet.

Example 1.88
Problem
Convert 40 psi to feet head.

Solution

$$\frac{40 \text{ psi}}{1} \times \frac{\text{ft}}{0.433 \text{ psi}} = 92.4 \text{ feet}$$

Example 1.89
Problem
Convert 40 feet to psi.

Solution

$$40 \frac{\text{ft}}{1} \times \frac{0.433 \text{ psi}}{1 \text{ ft}} = 17.32 \text{ psi}$$

As the above examples demonstrate, when attempting to convert psi to feet, we divide by 0.433, and when attempting to convert feet to psi, we multiply by 0.433. The above process can be most helpful in clearing up the confusion on whether to multiply or divide. There is another way, however—one that may be more beneficial and easier for many operators to use. Notice that the relationship between psi and feet is almost two to one. It takes slightly more than 2 feet to make 1 psi. Therefore, when looking at a problem where the data is in pressure and the result should be in feet, the answer will be at least twice as large as the starting number. For instance, if the pressure were 25 psi, we intuitively know that the head is over 50 feet. Therefore, we must divide by 0.433 to obtain the correct answer.

Example 1.90
Problem
Convert a pressure of 45 psi to feet of head.

Solution

$$45 \frac{\text{psi}}{1} \times \frac{1 \text{ ft}}{0.433 \text{ psi}} = 104 \text{ ft}$$

Example 1.91
Problem
Convert 15 psi to feet.

Solution

$$15 \frac{\text{psi}}{1} \times \frac{1 \text{ ft}}{0.433 \text{ psi}} = 34.6 \text{ ft}$$

Example 1.92
Problem
Between the top of a reservoir and the watering point, the elevation is 125 feet. What will the static pressure be at the watering point?

Solution

$$125 \frac{\text{psi}}{1} \times \frac{1 \text{ ft}}{0.433 \text{ psi}} = 54.1 \text{ ft}$$

Example 1.93
Problem
Find the pressure (psi) in a 12-foot deep tank at a point 5 feet below the water surface.

Solution

$$\text{Pressure (psi)} = 0.433 \times 5 \text{ ft} = 2.17 \text{ psi}$$

Example 1.94
Problem
A pressure gauge at the bottom of a tank reads 12.2 psi. How deep is the water in the tank?

Solution

$$\text{Head (ft)} = 2.31 \times 12.2 \text{ psi}$$

$$28.2 \text{ ft}$$

Example 1.95
Problem
What is the pressure (static pressure) 4 miles beneath the ocean surface?

Solution
Change miles to ft, then to psi.

$$5,280 \text{ ft/mile} \times 4 = 21,120 \text{ ft}$$

$$\frac{21,120 \text{ ft}}{2.31 \text{ ft/psi}} = 9,143 \text{ psi}$$

Example 1.96
Problem
A 150-ft diameter cylindrical tank contains 2.0 MG water. What is the water depth? What pressure would a gauge at the bottom read in psi?

Solution

Step 1: Change MG to cu ft.

$$\frac{2,000,000 \text{ gal}}{7.48} = 267,380 \text{ cu ft}$$

Step 2: Using volume, solve for depth.

$$\text{Volume} = .785 \times D^2 \times \text{depth}$$
$$267,380 \text{ cu ft} = .785 \times (150)^2 \times \text{depth}$$
$$\text{Depth} = 15.1 \text{ ft}$$

Example 1.97
Problem
The pressure in a pipe is 70 psi. What is the pressure in feet of water? What is the pressure in psf?

Solution

Step 1: Convert pressure to feet of water.

$$70 \text{ psi} \times 2.31 \text{ ft/psi} = 161.7 \text{ ft of water}$$

Step 2: Convert psi to psf.

$$70 \text{ psi} \times 144 \text{ sq in/sq ft} = 10,080 \text{ psf}$$

Example 1.98
Problem
The pressure in a pipeline is 6,476 psf. What is the head on the pipe?

$$\text{Head on pipe } = \text{ ft of pressure}$$
$$\text{Pressure } = \text{ Weight} \times \text{Height}$$
$$6{,}476 \text{ psf } = 62.4 \text{ lbs/cu ft} \times \text{ height}$$
$$\text{Height (ft of pressure) } = 104 \text{ ft}$$

Problems

1.1.   Convert a height of 6 feet 4 inches to a height in meters and centimeters.

1.2.   Determine the surface area of the moon (assuming it is a sphere) from its radius, which is about 1,738 km.

1.3.   Find the significant figure in each of the following numbers: 5 mL, 5.2 g, 5.0 kg, 5,000 L, and 0.005 m.

1.4.   Write the following numbers in scientific notation: (a) 1; (b) 30; (c) 5,720,000,000; (d) -0.0000000061.

1.5.   How many centimeters are in a millimeter?

1.6.   How many millimeters are in a kilometer?

1.7.   An object has a volume of 15 cm^3 and a mass of 45 g. What is its density?

1.8.   Another object has a volume of 30 cm^3 and a mass of 60 g. What is its density?

1.9.   A granite sample has a density of 2.8 g/cm^3. The density of water is 1.0 g/cm^3. What is the specific gravity of the granite?

1.10.   Another sample has a density of 174.8 lbs/ft^3. The density of water is 62.4 lbs/ft^3. What is the specific gravity of the sample now?

CHAPTER 2

Force and Motion

I can calculate the movement of the stars, but not the madness of men.

—Sir Isaac Newton

Every one of the measurable quantities that we discuss in this text can be specified in terms of only four basic dimensions: mass, length, time, and electric charge. In the following, we will begin a study of the first three of these.

In physics we are interested in trying to understand the motion of objects. In starting our discussion of motion we describe position in two and three dimensions and discuss the motion of objects in only one dimension. Later, after presenting and explaining foundational material, we discuss motion in regard to the real world; that is, we discuss the motion of objects in two dimensions (and in three dimensions, for that matter).

Position

Position and time are two fundamental quantities that can be used to describe where an object is, where it is headed, and how long it will take to get there. The position of an object along a straight line (its location in space, often indicated by the letter x; see figure 2.1) can be uniquely identified (we can measure it) by its distance from an origin (see figure 2.1). Because the position shown in figure 2.1 is specified by one

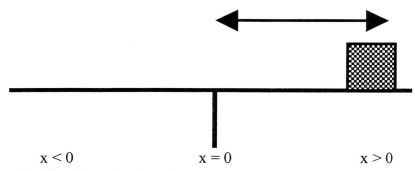

$x < 0$ $x = 0$ $x > 0$

Figure 2.1. One-dimensional position.

coordinate, it is said to be a one-dimensional problem. Some examples of one-dimensional motion include a car moving on a straight road, dropping a pencil, and throwing a ball straight up.

Speed and Velocity

In routine conversation, the terms *speed* and *velocity* are often used interchangeably—they are commonly thought to have the same meaning. In physics, however, they are two distinct quantities. *Speed* is a scalar quantity that refers to how fast an object is moving (i.e., the rate of change of distance with time). Thus, if you travel 16 miles in 2 hours, the average speed is 8 miles per hour (mph). Even though most people are confused about the difference between speed and velocity (defined below), most people do know the difference between two identical objects traveling at different speeds. Many know, for example, that a person moving faster (the one with the greater speed) will go farther in the same amount of time than one moving slower. If we do not understand this concept, we certainly know that someone moving faster will get to the same destination before someone who is moving slower. Intuitively, we all know that speed deals with both distance and time. We know that *sooner* means less time and that *faster* means greater distance (farther). Thus, it logically follows that to double one's speed means doubling one's distance traveled in a given amount of time. Moreover, doubling one's speed would also mean halving the time required to travel a given distance.

We can summarize our introductory presentation about speed by the following: A fast-moving object has a high speed and covers a relatively large distance in a short amount of time. A slow-moving object has a low speed and covers a relatively small amount of distance in a short amount of time. An object with no movement at all has a zero speed.

▶ **Key Point**: Simply, how fast an object moves is its speed.

Speed can be measured in meters per second, feet per second (ft/s), miles per hour, or knots. (A *knot* is 1 nautical mile per hour, or 1.15 statute miles per hour.) The standard formula for determining speed is

$$\text{speed} = \text{distance} \div \text{time} \tag{2.1}$$

or

$$s = d/t$$

Example 2.1
Problem
A truck leaves Norfolk, Virginia, at 11:00 a.m. EDT and arrives in Washington, D.C. (200 miles away) at 2:00 p.m. EDT the same day. What is the truck's average speed?

Solution

$$s = d/t$$

$$200 \text{ miles}/3 \text{ hours } + \sim 67 \text{ mph}$$

Velocity is speed with direction. Velocity is a vector quantity that refers to the rate at which an object changes its position. In regard to determining velocity, it is important to know (for example) whether you are traveling 50 mph due east or 50 mph due south. These same speeds (but with different velocities) will take you to two very different final locations. Problems in physics generally involve velocities (represented by vectors, which are simply measured quantities that have both a direction and magnitude [size]), because the direction of motion is typically an important piece of information.

Again, velocity is a vector quantity, and speed is a scalar quantity. Scalar quantities have magnitude, but no direction. Velocity is defined as a change in distance in a given direction divided by a change in time. The Greek letter delta (Δ) is used to represent the concept of change. Thus, the equation

$$\text{velocity} = \Delta x / \Delta t \tag{2.2}$$

is read as "velocity equals the change in x divided by the change in t."

To determine how far an object has traveled from an initial position, after a set amount of time, traveling at constant velocity, we need to derive equation (2.3) from equation (2.2):

$$x = x_o + vt \tag{2.3}$$

where

$$
\begin{aligned}
x &= \text{distance traveled} \\
x_o &= \text{initial position} \\
t &= \text{a set amount of time} \\
v &= \text{constant velocity}
\end{aligned}
$$

Example 2.2
Problem

You travel with an average velocity of 60 mph, you drive for 12 hours on your second travel day, and your starting point (x_o) was 400 miles beyond where you were the day before. Your total distance traveled at the end of the second day would be:

Solution

$$
\begin{aligned}
x &= (400 \text{ mi}) + (60 \text{ mph}) \times (12 \text{ h}) \\
x &= 1{,}120 \text{ mi}
\end{aligned}
$$

In typical physics problems involving the velocity of an object, there are sometimes several velocities involved, and the velocity that results from the sum of all the velocities described (the *resultant velocity*) must be determined. The resultant velocity is the sum of all the velocity vectors. A *vector* is a quantity that has direction as well as magnitude (size). The resultant velocity is the sum of all the velocity vectors.

To gain understanding of vectors and vector math operations, imagine a speedboat crossing a river. If the speedboat were to point its bow straight toward the other side of the river, it would not reach the shore directly across from its starting point (see figure 2.2). The river current influences the motion of the boat and carries it downstream. The speedboat may be moving with a velocity of 5 m/s directly across the river, yet the resultant velocity of the boat will be greater than 5 m/s and at an angle in the downstream direction. While the speedometer of the speedboat may read 5 m/s, its speed with respect to observers on the shore will be greater than 5 m/s. The resultant velocity of the boat can be determined by using vectors; it is the vector sum of the boat velocity and the river velocity. Because the speedboat heads across the river and because the current is always directed straight downstream, the two vectors are at right angles to each other. The lengths of the sides of a right triangle are related by the Pythagorean theorem (see figure 2.3), which states that in a right triangle the square of the length of the long side (hypotenuse) is equal to the sum of the squares of the other two sides. Generally, the Pythagorean theorem is written as $a^2 + b^2 = c^2$.

In our example of the speedboat crossing the river, the two velocity vectors form a right triangle, so that the resultant velocity can be computed with the formula

$$v \text{ (resultant)} = (v_1{}^2 + v_2{}^2)^{1/2}$$

where v_1 is the velocity of the river, and v_2 is the velocity of the speedboat.

Example 2.3
Problem

Suppose that the river in our example above is moving with a velocity of 4 m/s north, and the speedboat is moving with a velocity of 5 m/s east. What will be the resultant

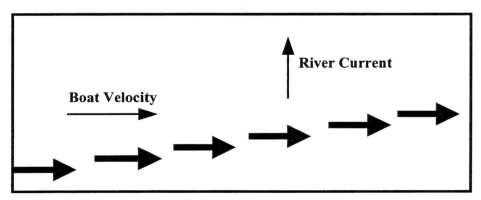

Figure 2.2.   Motion of speedboat with current.

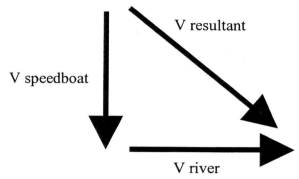

Figure 2.3. The lengths of the sides of a right triangle are related according to the Pythagorean theorem.

velocity of the speedboat (i.e., the velocity relative to observers on the shore)? The magnitude of the resultant can be found as follows:

Solution

$$(5.0 \text{ m/s})^2 + (4.0 \text{ m/s})^2 = R^2$$
$$25 \text{ m}^2/\text{s}^2 + 16 \text{ m}^2/\text{s}^2 = R^2$$
$$41 \text{ m}^2/\text{s}^2 = R^2$$
$$\sqrt{(41 \text{ m}^2/\text{s}^2)} = R$$
$$6.4 \text{ m/s} = R$$

Acceleration

When an object's velocity increases, we say it accelerates. Acceleration shows the change in velocity in a unit of time, or

$$\text{acceleration} = \Delta v/\Delta t$$

Because velocity is measured in meters per second (m/s) and time is measured in seconds (s), acceleration is measured in (m/s)/s, or m/s², which can be both positive and negative and always varying; it is rarely constant.

Accelerations determine the final velocity of an object. To determine an object's final velocity (v), given that it started at some initial velocity (v_o) and experienced acceleration (a) over a period of time (t), use the equation derived from the definition of acceleration:

$$v = v_o + at$$

Example 2.4
Problem
A truck starts at rest (with an initial velocity $v_o = 0$) and it accelerates for 5 s at an acceleration rate of 12 m/s², then what will be the final velocity?

Solution

$$v = v_o + at$$
$$v = 0 \text{ m/s} + (12 \text{ m/s}^2) \times (5 \text{ s})$$
$$v = 60 \text{ m/s}$$

It is important to note that a change in direction also constitutes acceleration. Remember, our definition of *velocity* is speed in a given direction. Thus, a change in direction is a change in velocity, and any change in velocity is acceleration. Moreover, whenever we step on the brakes to slow our cars we are experiencing another kind of acceleration called *deceleration*—a negative acceleration, or slowing down.

THE ACCELERATION OF GRAVITY

Downward on Earth a free-falling object has an acceleration of 9.8 m/s/s. The acceleration of gravity at the surface of Earth is often referred to as 1 g. Any object that is falling to the surface of Earth owing to the acceleration of gravity is in *free fall*. When the velocity and time for a free-falling object being dropped from a position of rest is tabulated, it displays the pattern shown in table 2.1.

The general acceleration equation is:

$$x = x_o + v_o t + \tfrac{1}{2} a t^2$$

where

$x =$ position of object
$v_o =$ initial velocity (original velocity)
$t =$ time elapsed
$a =$ constant acceleration

Example 2.5
Problem
A water balloon is dropped from a five-story building. How long will it take to hit the street below? Ignoring air friction, the only acceleration involved is the acceleration of gravity (g), and the height of the building is 20 m.

Table 2.1   Velocity and time for a free-falling object

Time (s)	Velocity (m/s)
0	0
1	−9.8
2	−9.6
3	−29.4
4	−39.2
5	−49.0

Solution

In the acceleration equation, we use $x = 20$ m (when the balloon hits the street), $x_o = 0$ m (we take the balloon's starting point at the top of the building to be zero), $v_o = 0$ m/s (the balloon starts from rest), and $g = 9.8$ m/s² (only gravity is acting on the balloon). Inserting these values into the equation and solving for t, we determine that

$$x = x_o + v_o t + \tfrac{1}{2}at^2$$
$$20 \text{ m} = 0 \text{ m} + 0 \text{ m/s}(t) + \tfrac{1}{2}(9.8 \text{ m/s})t^2$$
$$40/9.8 \text{ s}^2 = t^2$$
$$t = 2.0 \text{ s}$$

Force

In physics, we define *force* as a push or pull from the object's interaction with another object that can cause an object with mass to accelerate. Force has both magnitude (size) and direction, making it a vector quantity. When the interaction between two objects ceases, the objects no longer experience the force. All interactions (forces) between objects can be placed into two categories: contact forces (e.g., friction, normal, applied forces, etc.) and forces resulting from action-at-distance (e.g., magnetic, gravitational, or electrical force). Force is represented in units of newtons (abbreviated N). A *newton* is the force required to accelerate a 1-kg mass at a rate of 1 m/s².

Newton's Laws of Motion

To this point in the text, we have presented foundational information important to grasping the concepts put forward by Isaac Newton in his three laws of motion, which changed our understanding of the universe. Through careful observation and experimentation, Newton was able to describe the motion of objects by what are now called Newton's laws of motion. These laws include the *law of inertia* (Newton's first law of motion), the *law of constant acceleration* (Newton's second law of motion), and the *law of momentum* (Newton's third law of motion).

NEWTON'S FIRST LAW

> An object at rest will remain at rest unless acted on by an unbalanced force. An object in motion continues in motion with the same speed and in the same direction unless acted upon by an unbalanced force.

Basically, this law expresses what we mean when we say that an object has inertia. *Inertia* is that property of matter that causes matter to resist change in motion—it is the natural tendency for objects to keep on doing what they're doing.

✔ Note: *Mass* is a measure of how much inertia an object possesses.

NEWTON'S SECOND LAW

> Acceleration is produced when a force acts on a mass. The greater the mass (of the object being accelerated), the greater the amount of force needed (to accelerate the object).

Basically, we all know that heavier objects require more force to move the same distance as lighter objects. The second law does give us, however, an exact relationship between force, mass, and acceleration. It can be expressed as a mathematical equation:

$$F = MA$$

or

$$\text{force} = \text{mass times acceleration} = \text{Newtons} (\approx 1/4 \text{ pound})$$

Example 2.6
Problem
A four-wheeled cart weighs 1,200 kg. It is at rest. A man tries to push the cart to a storeroom, and he makes the cart go 0.05 m/s/s. How much force is the man applying to the cart?

Solution

$$F = MA$$
$$F = 1200 \times 0.05$$
$$F = 60 \text{ newtons}$$

NEWTON'S THIRD LAW

> For every action there is an equal and opposite reaction.

Basically, what this means is that whenever an object pushes another object, it gets pushed back in the opposite direction equally hard.

Problems

2.1. You use a U.S. atlas calculator to calculate the driving distance from your house to your friend's house. The results say that the distance is 375 mi, and the travel time is 7 h. What speed is the atlas mileage calculator assuming you will average for the trip?

2.2. A skydiver reaches a terminal velocity of 15 mi/h at a height of 1 mi above Earth. How long will it take him or her to reach the ground (in minutes)?

2.3. A jet plane flies due south at 520 mph, in a 50 mph crosswind, blowing from the west. What is the resultant velocity (speed and direction) of the plane?

2.4. A rocket accelerates from rest at 1 g (9.8 m/s²) for 12 min. What is the final velocity of the rocket after this 12-min acceleration (in km/s)?

2.5. A bag of garbage is dropped from a cliff 60 m above an ocean beach. Assuming the bag never reaches terminal velocity, how many seconds will it take for the bag to hit the beach?

2.6. Like velocity, force has both a(n) _____ and a(n) _____.

2.7. Which of Newton's laws says that for every action there is an equal and opposite reaction?

CHAPTER 3

Work, Energy, and Momentum

> Physics does not change the nature of the world it studies, and no science
> of behavior can change the essential nature of man, even though both sci-
> ences yield technologies with a vast power to manipulate their subject mat-
> ters.
>
> —B. F. Skinner

Work

In physics, *work* is defined as the product of the net force and the displacement
through which that force is exerted, or $W = F$ (Force) $\times d$ (displacement). (F and d
are vectors but W is not.) This corresponds to our everyday meaning of the word in
that when you lift a grocery bag 4 feet from the floor you do twice as much work as
when you lift it 2 feet. Or suppose you lift 10 grocery bags to a height of 3 feet. You
do 30 times as much work as lifting one grocery bag that high. Keep in mind that
although work is a scalar quantity, both F and d are vector quantities that must be in
the same direction if they are to be multiplied to obtain work.

When force is measured in pounds and distance in feet, the unit of work is the
foot-pound (ft-lb). In SI units, recall that force is measured in newtons and distance in
meters. The unit of work is the newton-meter, called the *joule* (pronounced "jewel").

To gain a better understanding of the information provided above, consider the
forces acting on the box shown in figure 3.1. The forces acting are represented by
arrows, and also show the acceleration of the box as an arrow.

As shown in figure 3.1, different forces act on the box.

- W is the weight of the box
- N is the force on the box arising from contact with the floor
- F is the constant force applied to the box
- a represents the acceleration of the box

Example 3.1
Problem:
A force $F = 20\ N$ pushes a container across a frictionless floor for a distance $d = 10$
m. Determine how much work is done on the box.

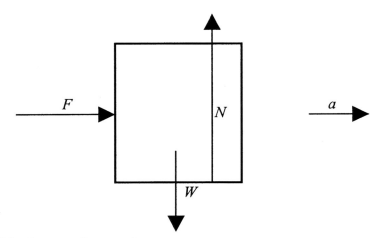

Figure 3.1.   Forces acting on a box.

Solution:

$$W = F \times d$$
$$= (20 \text{ N}) \times (10 \text{ m}) = 200 \text{ joules (J)}$$

Energy

Energy (often defined as the ability to do work) is one of the today's most-discussed topics because of high prices for hydrocarbon products (gasoline and diesel fuel), electricity, and natural gas. These are forms of energy that we are quite familiar with, but energy also comes in other forms—thermal (heat), radiant (light), mechanical, and nuclear energy. Energy is in everything. All things we do in life and death (biodegradation requires energy, too) are a result of energy. There are two types of energy—stored (potential) energy, and working or moving (kinetic) energy.

POTENTIAL ENERGY

An object can have the ability to do work—have energy—because of position. For example, a weight suspended high from a scaffold can be made to exert a force when it falls. Because gravity is the ultimate source of this energy, it is correctly called gravitational potential energy or GPE (GPE = weight × height), but we usually refer to this as potential energy or PE.

Another type of potential energy is chemical potential energy—the energy stored in a battery or the gas in a vehicle's gas tank.

Consider figure 3.2. When the suspended object is released, it will fall on top of

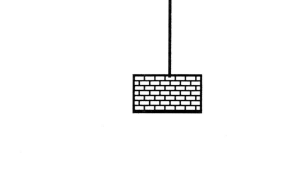

Figure 3.2.   Gravitational potential energy (GPE): a box of bricks suspended above an empty cardboard box.

the box and crush it, exerting a force on the box over a distance. By multiplying the force exerted on the box by the distance the object falls, we can calculate the amount of work that is done.

KINETIC ENERGY

Moving objects have energy (the ability to do work). The kinetic energy of an object is related to its motion. Figure 3.3 shows the suspended box of bricks we used earlier to demonstrate potential energy, but now the box of bricks is free-falling—the potential energy is converted to kinetic energy because of movement. Specifically, the *kinetic energy* (KE) of an object is defined as half its mass times its velocity squared, or

$$KE = \tfrac{1}{2}\, mv^2$$

From this equation, it is apparent that the more massive an object and the faster it is moving, the more kinetic energy it possesses. The units of kinetic energy are determined by taking the product of the units for mass (kg) and velocity squared (m^2/s^2). The units of KE, like PE, are joules. Kinetic energy can never be negative and only tells us about speed, not velocity.

Momentum

An object in motion possesses *momentum*. Momentum, abbreviated with the letter *p*, is equal to the product of mass and velocity, $p = mv$. Thus, the SI unit of momentum

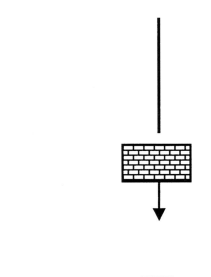

Figure 3.3.   Kinetic energy (KE): free-falling box of bricks suspended above an empty cardboard box.

is kg · m/s. Momentum has direction; therefore it is a vector quantity. Newton's third law—for every action there is an equal and opposite reaction—is often referred to as the *law of conservation of momentum*.

Using a standard, well-worn, but appropriate example of someone jumping from a rowboat into a body of water to swim and cool off, helps to explain momentum and the law of conservation. First we write total momentum as

$$p_{total} = 0 = m_1v_1 + m_2v_2$$

or

$$m_1v_1 = -m_2v_2$$

and then designating the two objects as the boat, m_1 and using m_2 for the swimmer, we observe that the two objects will move in opposite directions. If the mass of the boat (m_1) is 150 kg and the mass of the swimmer (m_2) is 50 kg, then the velocity of the boat (v_1) will be

$$v_1 = (-m_2/m_1) v_2$$

or

$$v_1 = -0.33 v_2$$

—indicating that the boat will move away from the swimmer at one-third the speed of the swimmer swimming away from the boat. Keep in mind that if some force other than the swimmer jumping from the boat had caused the boat to move, the principle of conservation of momentum would not apply. This is because the application of the principle of conservation of momentum is limited to systems isolated from other forces (basically, it must be a closed system).

Problems

3.1. If you double the mass or double the velocity of an object, which has a greater effect on its kinetic energy?

3.2. A man is pushing a desk across a wooden floor. A constant force of 250 N required, and he pushes the desk over a distance of 15 m. How much work does the man do?

3.3. Which has more potential energy, a 1000-kg rock suspended 10 m in the air, or a 4-kg bag of marbles suspended at 200 m? How much work was required to get each object from the ground to their higher potential energy positions?

3.4. You are pushing your 30-kg child on a playyard swing. If the child's height is 2.2 m at the maximum height of his or her swing, how fast will the child be going at the lowest point in his or her swing?

3.5. In the course of a 30-min workout session, you lift a total of 12,000 kg, an average of 60 cm. How much work did you do? How much power was required for your workout?

CHAPTER 4

Circular Motions and Gravity

Physics is becoming so unbelievably complex that it is taking longer and longer to train a physicist. It is taking so long, in fact, to train a physicist to the place where he understands the nature of physical problems that he is already too old to solve them.

—Eugene Paul Wigner

Circular motion, or more generally, *angular motion*, can include race cars whizzing around a track, rockets moving around planets, bees buzzing around a hive, children on a merry-go-round, a yo-yo, planets orbiting the sun, and the propeller of an airplane. In the previous chapters we discussed concepts like displacement, velocity, and acceleration and how they relate to linear motion; now we will discuss the terms needed to describe and predict angular motion. Before discussing angular motion, qualitatively and quantitatively (and to aid in understanding the material to follow), a comparison of linear and angular motion terms is presented in table 4.1.

Angular Motion

Earlier, when we discussed linear motion, we pointed out that linear distances are measured in meters and in other linear units of measure. These linear measurements were displacements from an arbitrary zero-point. Angular motion is measured differently. The different ways angular motion or angular distance is measured are shown in figure 4.1 and explained below.

Suppose the circular objects shown in figure 4.1 are axle-mounted wheels wherein the top wheel is rotated through θ as shown. This rotation can be measured in three

Table 4.1   Terms used for linear and angular motion

Linear motion	Units	Angular motion	Units
Distance (x)	m	Angle (θ)	rad°
Velocity (v)	m/s	Angular velocity (ω)	rad/s or 1/s
Acceleration (α)	m/s^2	Angular acceleration (α)	rad/s^2 or 1/s^2
Force (F)	N	Torque (τ)	N·m
F = ma	N	$\tau = I\alpha$	N·m
Mass (m)	kg	Moment of inertia (I)	kg·m^2

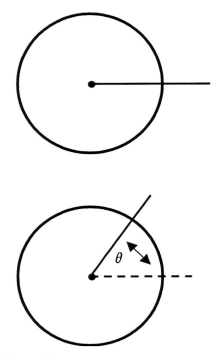

Figure 4.1. Angular distance θ.

different ways. As you might expect, the two most common ways, though arbitrary, to measure the angle (θ) are with revolutions (one full rotation is a revolution) and degree units (one full rotation is defined as 360°)—these are related by

$$1 \text{ rev } = 360°$$

The third way to measure θ is not arbitrary. It is an angular measure scaled to the circle, called a *radian* (rad). This is shown in figure 4.2.

Angular Velocity

Angular velocity is a measure of the rate of change in angular position. In an average day, we are exposed to many different examples of angular velocity. Consider, for example, the tachometer in your automobile. The automobile tachometer measures the number of revolutions per minute (rev/min) of the engine's crankshaft. When the auto is not moving (idling), the tachometer reads somewhere around 600 to 1,000 rev/min. Actually, the car engine's crankshaft is idling but not stopped; it is revolving at what is known as angular speed (the number of revolutions per minute that the crankshaft is making). When you step on the gas pedal (accelerator), you feed more

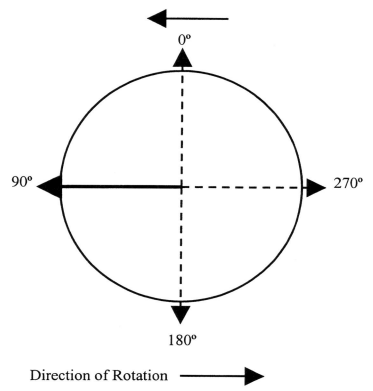

Figure 4.2.   Degrees and radians in circular motion: 90° = π/2 radians; 360° = 2π radians.

gas to the engine, causing the crankshaft to rotate at a higher angular velocity—the motion of the crankshaft is transferred to the wheels of the car manually or automatically by the transmission.

The average angular velocity of a rotating object is defined to be the angular distance divided by the time taken to turn through this angle, as defined in the following equation:

$$\text{average angular velocity } (\omega_{AVE}) = \theta/t$$

where θ is the angle through which an object rotates in time t. The units for angular velocity (ω, the Greek letter "omega") are those of an angle divided by a time—units might be degrees per second, revolutions per minute, or radians per second.

Angular Acceleration

When a rotating object's angular velocity changes—it speeds up or slows down—it signifies the presence of angular acceleration. Because angular accelerations are changes in velocity, the units of angular acceleration are radians per second per second. Because radians are dimensionless units, angular acceleration is measured in units of $1/s^2$. Like angular velocity, ω, *angular acceleration* (α, the Greek letter "alpha") is a vector, meaning it has a magnitude and a direction. Angular acceleration is the rate of change of angular speed:

$$\alpha = \Delta\omega \,/\, \Delta t$$

Torque

Torque is a vector that measures the tendency of a force to rotate (or twist) an object about some axis. The SI unit for torque is Newton-meters (Nm). In U.S. customary units, it is measured in foot pounds (ft-lb). The symbol for torque is τ, the Greek letter *tau*. In physics, torque can be described simply as the application of a force at some distance from a pivot point (see figure 4.3), or

$$\text{torque} = \text{force} \times \text{radius}$$

Angular Momentum

The *angular momentum* of an object around a certain point is a vector quantity equal to the product of the distance from a point and its momentum measured with respect to the point. In equation form, we say that an object with a mass m moving in a circle of radius r at a velocity v has an angular momentum of

$$L = mvr$$

Gravity

Gravity refers to the force that is the cause of the attraction between objects. Newton, using foundational scientific explanations provided earlier by Kepler and Brahe, was

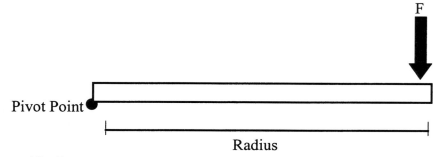

Figure 4.3.   Torque.

the first to describe the force that makes the planets move in elliptical paths around the sun. Newton claimed that the same physical laws that can explain how things move here on Earth can explain the motions of the planets. Newton asserted that all objects with mass exert an attractive force on all other objects with mass. Newton further posited that this force is directly proportional to the product of the masses of the two objects and inversely proportional to their distance squared.

Newton's law of gravity can be stated as follows:

$$F \sim \frac{m_1 \times m_2}{d^2}$$

The symbol $\sim$ means "is proportional to"; m_1 symbolizes the mass of one of the objects, and m_2 is the mass of the other; d is the distance between the centers of the objects.

Problems

4.1. If you increase the mass of an object two times, but it is still the same size, how would the weight differ?

4.2. A spinning top is rotating once every 1 s. How many degrees does it rotate through in half a revolution? How many radians? After 200 rev, how many degrees has it rotated through? How many radians?

4.3. Before there were CDs, there were vinyl records that rotated on a turntable at 45 rev/min. What was their rotation rate in radians per second?

4.4. The spinning top in problem 4.2 started at rest. If it took 5 s to get it to a rotation rate of once every 5 s, what angular acceleration did the spinning top undergo?

4.5. As one rises above the surface of Earth, does the acceleration of gravity become greater or less?

CHAPTER 5

Atoms, Molecules, and Elements

In science, "fact" can only mean confirmed to such a degree that it would be perverse to withhold provisional assent. I suppose that apples might start to rise tomorrow, but the possibility does not merit equal time in physics classes.

—Stephen Jay Gould

All matter is composed of atoms. Our bodies, the clothes we wear, the cars we drive and the boats we sail, the houses we live in—all these objects are composed of atoms. There are three states of matter—solid, liquid, and gas. Atoms combine to form molecules. Matter is held together by attractive forces—forces that prevent substances from coming apart. The molecules of a solid are packed more closely together and have little freedom of motion. In liquids, molecules move with more freedom and are able to flow. The molecules of gases have the greatest degree of freedom and their attractive forces are unable to hold them together.

The kinetic theory of matter (*kinetic* is the Greek word that means "motion"; *theory* is the Greek word for "idea") is a statement of how we believe atoms and molecules behave and how it relates to the ways we have to look at the things around us. Essentially, the theory states that all molecules are always moving. More specifically, the theory says:

- All matter is made of atoms, the smallest bit of each element. A particle of a gas could be an atom or a group of atoms.
- Atoms have an energy of motion that we feel as temperature. At higher temperatures, the molecules move faster.

Matter changes its state from one form to another. Examples of how matter changes its state include the following:

- Melting is the change of a solid into its liquid state.
- Freezing is the change of a liquid into its solid state.
- Condensation is the change of a gas into its liquid form.
- Evaporation is the change of a liquid into its gaseous state.
- Sublimation is the change of a gas into its solid state and vice versa (without becoming liquid).

Key Definitions

- *Matter*: anything that takes up space and has mass.
- *Element*: a substance that cannot be broken down by ordinary means; the material making up matter.
- *Atoms*: small units of matter. *Protons* $(+)$, *neutrons* (0), and *electrons* $(-)$ are the subunits of atoms.
- *Atomic number*: the total number of protons in an atom.
- *Atomic mass*: the total number of neutrons and protons in an atom.
- *Isotopes*: different atomic forms caused by varying the number of electrons.
- *Energy levels*: All electrons have the same mass and charge. They differ in the amounts of potential energy they possess. Electrons closer to the nucleus contain less potential energy.
- *The periodic table*: a chart made up of eighteen columns and seven rows. The elements on the table are organized by increasing atomic number. The periodicity of the elements defines unique properties such that elements of a given column in the table have similar properties.

Atomic Theory

The ancient Greek philosophers believed that all the matter in the universe was composed of four elements: earth, fire, air, and water. Today we know of more than 100 different elements, and scientists are attempting to create additional elements in their laboratories. An *element* is a substance from which no other material can be obtained. Today we also know that atoms are the basic building blocks of all *matter*. Atoms equal the smallest particle of matter with constant properties. Atoms are so small that it would take approximately *two thousand million* atoms side by side to equal one meter in length! Atoms are so small that scientists were forced to devise special weights and measures:

- Mass/weight: *atomic units (au)*
  - 1 au $= 1.6604 \times 10^{-24}$ g
- Length: *angstrom (Å)*
  - 1 Å $= 10^{-8}$ cm

It is interesting to note that scientists used to believe that atoms were *indivisible*, but we now know that they are made up of many subatomic particles. Physics primarily deals with three subatomic particles: protons, neutrons, and electrons.

- *Protons* and *neutrons* are located in the nucleus (center) of the atom (see figure 5.1).
  - Protons are positively charged particles weighing 1 atomic unit.
  - Neutrons have no charge. They weigh 1 atomic unit.
- *Electrons* are located in "orbitals" around the nucleus (see figure 5.1).
  - Electrons are negatively charged particles that have no mass.

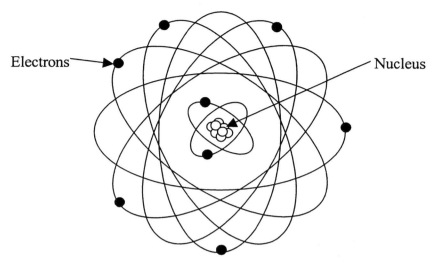

Electrons   Nucleus

Figure 5.1. Electrons and nucleus (protons and neutrons) of an atom.

The *stability* of a nucleus depends on the balance between attractive gravitational forces, repulsive electronic forces, and the ratio of protons to neutrons. In a stable atom (or neutral atom), the number of electrons equals the number of protons.

ELEMENTS

As mentioned, scientists have identified more than 100 different types of atoms, which they call *elements*. An atom is the smallest unit of an element that still retains the properties of that element.

✔ **Key Point:** An *element* is a substance from which no other material can be obtained. Stated differently, the chemical elements are the simplest substances into which ordinary matter may be divided. All other materials have more complex structures and are formed by the combination of two or more of these elements.

Elements are substances; therefore they have physical properties that include density, hardness, odor, color, and so on, and chemical properties that describe the ability to form new substances (i.e., a list of all the chemical reactions of that material under specific conditions).

Each element is represented by a *chemical symbol*. Chemical symbols are usually derived from the element's name (e.g., Al for aluminum). The chemical symbols for the *elements* known in antiquity are taken from their Latin names (e.g., Pb for lead). For every element there is one and only one uppercase letter (e.g., O for oxygen). There may or may not be a lowercase letter with it (e.g., Cu for copper). When written in chemical equations, we represent the elements by the symbol alone with no charge attached.

EXAMPLES OF CHEMICAL SYMBOLS

Fe (iron)	P (phosphorus)
Al (aluminum)	Ag (silver)
Ca (calcium)	Cl (chlorine)
C (carbon)	Cu (copper)
N (nitrogen)	K (potassium)
Rn (radon)	He (helium)
H (hydrogen)	Si (silicon)
Cd (cadmium)	U (uranium)

✔ **Key Point:** Presently, we know only 100+ elements, but well over a million compounds. Only 88 of the 100+ elements are present in detectable amounts on Earth, and many of these 88 are rare. Only 10 elements make up approximately 99% (by mass) of the Earth's crust, including the surface layer, the atmosphere, and bodies of water (see table 5.1).

As can be seen from table 5.1, the most abundant element on Earth is oxygen, which is found in the free state in the atmosphere as well as in combined form with other elements in numerous minerals and ores.

MATTER AND ATOMS

Molecules consist of two or more atoms. When atoms (elements) chemically combine to form compounds, they lose all their original properties and create a new set of properties unique to the compound. For example, sodium (Na) and chlorine (Cl) are poisonous, but they combine to form a compound called sodium chloride, which is ordinary table salt.

As mentioned, a molecule is the smallest unit of a compound that still retains the

Table 5.1   Elements making up 99% of Earth's crust, oceans, and atmosphere

Element	Symbol	% composition	Atomic number
Oxygen	O	49.5%	8
Silicon	Si	25.7%	14
Aluminum	Al	7.5%	13
Iron	Fe	4.7%	26
Calcium	Ca	3.4%	20
Sodium	Na	2.6%	11
Potassium	K	2.4%	19
Magnesium	Mg	1.9%	12
Hydrogen	H	1.9%	1
Titanium	Ti	0.58%	22

properties of that compound. *Compounds* are two or more elements that are "stuck" (bonded) together in definite proportions by a chemical reaction—for example, water (H_2O) and halite or table salt (NaCl). Compounds always have uniform proportions, and they have unique properties that are different from their components and cannot be separated by physical means. Each individual compound will always contain the same elements in the same proportions by weight.

$$H_2O \text{ (water)} \neq H_2O_2 \text{ (hydrogen peroxide)}$$

Compounds are represented by *chemical formulas*. Chemical formulas consist of *chemical symbols* and subscripts to describe the relative number of atoms present in each compound.

Common chemical formulas include

- H_2O (water)
- NaCl (sodium chloride, table salt)
- HCl (hydrochloric acid)
- CCl_4 (carbon tetrachloride)
- CH_2Cl_2 (methylene chloride)

🖒 **Key Point:** A chemical formula tells us how many atoms of each element are in the molecule of any substance. As mentioned, the chemical formula for water is H_2O. The "H" is the symbol for hydrogen. Hydrogen is a part of the water molecule. The "O" means that oxygen is also part of the water molecule. The "$_2$" after the H means that two atoms of hydrogen are combined with one atom of oxygen in each water molecule.

Most naturally occurring matter consists of mixtures of elements and/or compounds. Mixtures are found in rocks, the ocean, vegetation, just about anything we find. *Mixtures* are combinations of elements and/or compounds held together by physical rather than chemical means. It is important to note that mixtures are physical combinations and compounds are chemical combinations.

Mixtures have a wide variety of compositions. Mixtures can be separated into their ingredients by physical means (e.g., filtering, sorting, distillation, etc.). The components of a mixture retain their own properties.

🖒 **Key Point:** The thing to remember about mixtures is that we start with some pieces, combine them, and we can do something to pull those pieces apart again. We wind up with the same molecules (in the same amounts) that we started with. Two classic examples of mixtures are concrete and salt water. When mixed, they seem to form compounds, but because physical forces (cement can still have the basic parts removed by grinding and salt water by filtering), the parts are just like when they were when we started.

Electron Configuration

Recall that an atom contains subatomic particles including:

- protons ($+$), positively charged particles
- electrons ($-$), negatively charged particles
- neutrons (0), no charge

Free (unattached), uncharged atoms have the same number of electrons as protons to be electrically neutral. The protons are in the nucleus and do not change or vary except in some nuclear reactions. The electrons are in discrete pathways or shells (orbits) around the nucleus. There is a ranking or hierarchy of the shells, usually with the shells further from the nucleus having a higher energy.

In an atom, the electrons seek out the orbits that are closest to the nucleus, because they are located at a lower energy level. The low energy orbits fill first. The higher energy levels fill with electrons only after the lower energy levels are occupied. The lowest energy orbit, which is closest to the nucleus, is labeled the *K shell*. The outer orbits, or shells, are listed in alphabetical order: K, L, M, N, O, P, and Q.

In the atomic diagram shown (figure 5.2), it can be seen that 2 electrons are needed to fill the K shell, 8 electrons to fill the L shell, and, for light elements (atomic numbers 1–20), 8 electrons will fill the M shell.

Electron configuration is the "shape" of the electrons around the atom—that is, which energy level (shell) and what kind of orbital it is in. The shells were historically named for the chemists who found and calculated the existence of the first (inner) shells. Their names began with "K" for the first shell, then "L," then "M," so subsequent energy levels were continued up the alphabet (see figure 5.2). The numbers 1 through 7 have since been substituted for the letters.

The electron configuration is written out with the first (large) number as the shell number. The letter is the orbital type (either *s, p, d,* or *f*). The smaller superscript number is the number of electrons in that orbital.

K	L	M	N	O	P	Q	
1	2	3	4	5	6	7	
s	*sp*	*spd*	*spdf*	*spdf*	*spd*	*sp*	
2	8	8	2				20
		10	6	2			38
			10	6	2		56
			14	10	6	2	88
				14	10	6	
2	8	18	32	32	18	8	TOTALS

Figure 5.2.   Electron configuration chart.

To use this scheme, you first must know the orbitals. An *s* orbital only has 2 electrons. A *p* orbital has 6 electrons. A *d* orbital has 10 electrons. An *f* orbital has 14 electrons. We can tell what type of orbital it is by the number on the chart. The only exception to that is that "8" on the chart is "2" plus "6"—that is, an *s* and a *p* orbital. The chart reads from left to right and then down to the next line, just as English writing. Any element with over 20 electrons in the electrical neutral unattached atom will have all the electrons in the first row on the chart.

The totals on the right indicate using whole rows. If an element has an atomic number over 38, take all the first two rows and whatever more from the third row. For example, iodine is number 53 on the periodic table (discussed later). For its electron configuration we would use all the electrons in the first two rows and 15 more electrons: 1*s*2 2*s*2 2*p*6 3*p*6 4*s*2 3*d*10 4*p*6 5*s*2 from the first two rows and 4*d*10 5*p*5 from the third row. We can add up the totals for each shell at the bottom. Full shells would give us the totals on the bottom.

✔ **Key Point:** Electron configuration is the "shape" of the electrons around an atom—that is, which energy level (shell) and what kind of orbital it is in.

Periodic Table of Elements

The *periodic table of elements* is an arrangement of the elements into rows and columns in which those elements with similar properties occur in the same column. Chemists use this table to correlate the chemical properties of known elements and predict the properties of new ones.

✔ **Key Point:** The periodic table of elements is a way to arrange the elements, based on electronic distribution, to show a large amount of information and organization.

The periodic table of elements provides information and organization on the following:

- Atomic number (the number of electrons revolving about the nucleus of the atom)
- Isotopes (atoms of an element with different atomic weights)
- Atomic weight and molecular weight
- Groups (vertical columns) and periods (horizontal columns)
- Locating important elements

Remember that each element is represented by a chemical symbol, for example:

Fe (iron)	P (phosphorus)
Al (aluminum)	Ag (silver)
Ca (calcium)	Cl (chlorine)
C (carbon)	H (hydrogen)

Atomic number is the *number of protons* in the nucleus of an atom (also the number of electrons surrounding the nucleus of an atom). Remember that *protons* are positively charged subatomic particles found in the nucleus of the atom.

✔ **Key Point:** The number of protons in the nucleus determines the atomic number. Changing the number of protons changes the element, the atomic number, and the atomic mass.

As mentioned, the atomic number can also indicate the *number of electrons* if the *charge* of the atom is known:

Element	Atomic No.	Charge	No. of Electrons
O	8	0	8
O^{-2}	8	-2	10
Na	11	0	11
Na^+	11	$+1$	10

The atomic number (again, the number of protons) is a unique identification number for each element. It cannot change without changing the identity of the element:

$$C = 6 \qquad O = 8 \qquad Cl = 17$$

The *number of neutrons* can change without changing the identity of the element or its chemical and physical properties. Atoms with the same number of protons but different numbers of neutrons are called *isotopes*:

Isotope	No. Protons	No. Neutrons
^1H (hydrogen)	1	0
^2H (deuterium)	1	1
^3H (tritium)	1	2

If the number of neutrons changes, the atomic weight of the atom changes:

Isotope	No. Protons	No. Neutrons	Atomic Weight
^1H (hydrogen)	1	0	1 au
^2H (deuterium)	1	1	2 au
^3H (tritium)	1	2	3 au

Atomic weight is the relative weight of an average atom of an element, based on C^{12} being exactly 12 atomic units. Chemists generally *round* atomic weights to the nearest whole number:

$$H = 1 \qquad O = 16 \qquad Fe = 56$$

Chemists add up atomic weights to determine the *molecular weight* of a compound. Molecular weights are an important indication of the general size (and therefore complexity) of a compound. In general (but not always), the higher the molecular weight, the greater the likelihood that the compound may persist in the environment.

Periods, the rows (horizontal lines) of the periodic table, are organized by increasing atomic number. *Groups* are the columns of the periodic table, which contain elements with similar chemical properties. Within each period or row, the *size* of the nuclei increases going from left to right because the *atomic number* increases.

Within each group or column, the *size* of the atoms increases going from top to bottom because the *number of electron shells* increases.

Within each group or column, elements have similar chemical reactivity because they have a *similar number of outer shell electrons.*

The *group numbers* (i.e., Roman numerals) above each column indicate the number of outer shell electrons in each group:

Group V = 5 outer shell electrons
Group II = 2 outer shell electrons
Group VII = 7 outer shell electrons

Only "outer shell" electrons are involved in chemical change. The nucleus and inner shell electrons are not altered in any way during ordinary chemical reactions.

Remember that the periodic table lists each element and its chemical symbol (e.g., Fe [iron], Al [aluminum], C [carbon], etc.). Along with the element's name and chemical symbol, the atomic number and atomic weight are also listed (e.g., the atomic number for Fe is 26; the atomic weight for Fe is 55.847).

Chemical Bonding

As mentioned, atoms of elements can combine to form molecules of compounds. Experimentation has shown that only certain combinations of atoms will bond together. The tendency of elements to link together to form compounds through a shift of electronic structure is known as *valence.* This linking process is accomplished through *valence electrons.* These electrons occupy the last energy level of an atom, which is where atoms come in contact with each other. It stands to reason that chemical bonds will occur here in any chemical reaction. The maximum number of valence electrons any atom can contain is 8. Any number less than 8 will allow that atom to act as a donor or recipient of electrons to become stable. Atoms that give electrons will become positive ions and have a positive (+) charge, while atoms that receive electrons will become negative ions and have a negative (−) charge.

ABOUT CHEMICAL BONDS

Atoms are linked or joined by chemical bonds. Only electrons are involved in the formation of chemical bonds between atoms. Only the outermost electrons (i.e., the outer shell electrons) are typically involved in bonding. Each bond consists of two electrons, one from each atom in the bond.

There are two different types of chemical bonds, depending on the type of atoms that are bonded together: covalent and ionic bonds.

A *covalent bond* results from the sharing of a pair of electrons between two covalent bonds:

• *Nonpolar covalent bond*: Here the valence electrons are shared equally, thus eliminating a positive and negative end on the molecule. An example of a nonpolar bond is $CHCl_4$. It is nonpolar, because although the four C-Cl bonds are all polar, the symmetry of their arrangement around the central C atom makes the overall molecule nonpolar.
• *Polar covalent bond*: Here the valence electrons are shared unequally, causing the molecule to develop a positive end (where the electrons spend less time) and a negative end (where the electrons spend more time). This has to do with the electronegativity of the atom. The more electronegative the atom the more it will hold on to the electrons. Oxygen is very electronegative. Hydrogen is not. If the difference in electronegativity between two atoms is sufficiently large, the shared electron pair will spend all of its time on the more electronegative atom, resulting in ionic bonding rather than covalent bonding. An example of a polar covalent bond is $CHCl_3$, because of the polar C-Cl bond (Cl is an electronegative atom).

✔ **Key Point:** A polar covalent bond has a positively charged end and a negatively charged end.

In covalent bonding, the sharing of one pair of electrons is called a *single bond*; the sharing of two pairs of electrons is called a *double bond* (e.g., carbon dioxide); and the sharing of three pairs of electrons is called a *triple bond* (e.g., acetylene). Double bonds are more reactive than single bonds, and compounds containing double bonds are somewhat more volatile than corresponding single-bonded molecules. Triple bonds are even more reactive than double bonds, and volatility in triple-bonded compounds is still greater.

An *ionic bond* results from the transfer of electrons from one atom to another. In an ionic bond, one atom "donates" one or more of its outermost electrons to another atom or atoms.

• The atom that gains the electrons becomes a negative ion, or *anion.*

✔ **Key Point:** Remember the mnemonic for anion, "*a negative ion.*"

• The atom that loses the electron becomes a positive ion, or *cation.*

✔ **Key Point:** Covalent bonding is a process similar to ionic bonding, but the electrons are shared rather than transferred.

ELECTRONEGATIVITY AND POLARITY

Electronegativity is a measure of the ability of an atom to attract its outermost electrons. The higher the electronegativity, the greater the atom's ability to attract other electrons. Highly electronegative atoms tend to form *ionic* or *polar covalent* bonds. In polar covalent bonds, the more electronegative atom "keeps" the electron pair more (i.e., takes a larger share of the electron density). In nonpolar covalent bonds (e.g., C-C or C-H), the electrons are shared equally.

✔ **Key Point:** Think of chemical bonding as a continuum from nonpolar covalent bonds to ionic bonds. Bond polarity depends on the *electronegativity* of the atoms involved, but overall molecular polarity depends on symmetry.

BOND STRENGTHS AND BOND ANGLES

Chemists measure the *strength* of a bond by determining how much energy is needed to break the bond. Examples of carbon (C) bond strengths are shown in table 5.2.

In addition to measuring bond strengths, chemists can also measure bond angles between atoms. *Bond angles* depend on the type of atoms bonded together. Chemists predict *bond angles* and other 3-D structures using computer programs.

STRUCTURAL FORMULAS

The overall 3-D shape of a molecule is represented by its *structural formula*, which shows how the atoms are attached. In chemical shorthand, the Cs and Hs are often not shown. Structural formulas give us the same information as the chemical formulas, but they also tell us how the atoms are bonded together.

ISOMERS

Isomers are defined as one of several organic compounds with the same chemical formula but different structural formulas and therefore different properties. As mentioned, isomers are different compounds that have different properties, although these differences are often slight. Structure makes the difference!

Table 5.2   Carbon bond strengths

Type of bond	Strength (kcal/mol)
C-N	70
C-Cl	79
C-O	84
C-H	99

✔ **Interesting Point:** "*Iso*" is a Greek word that means same or equal. *Mer* means unit or formula. Isomers are different compounds, even though they have the same formula. For example, n-propanol has a boiling point of 97.2°C, and iso-propanol has a boiling point of 82.5°C (see also table 5.3).

Table 5.3   Xylene isomers

Compound	Boiling point	Melting point
o-xylene	144°C	−25°C
m-xylene	139.3°C	−47.4°C
p-xylene	138°C	14°C

HYDROGEN BONDING

Hydrogen bonds are bonds formed when hydrogen that is covalently bonded to an electronegative atom is attracted to another electronegative atom on another molecule. Hydrogen bonds are *weak* attractions between *molecules*, whereas ionic and covalent bonds are much stronger attractions between *atoms*.

When H is bonded to O or N, its lone electron is pulled away from its single-proton nucleus. A broken line represents hydrogen bonds because they are much weaker than covalent bonds.

✔ **Key Point:** Hydrogen bonds are like "glue" between molecules.

Although a single hydrogen bond is very weak, the cumulative effect of an enormous number of hydrogen bonds can significantly affect the properties of the compound.

SHORT REVIEW

When an atom combines chemically with another atom, it will either:

1. Gain electrons (become a negatively charged ion, or anion)
2. Lose electrons (become a positively charged ion, or cation)
3. Share electrons

Problems

5.1. Atoms are composed of neutrons, protons, and electrons. What is the charge on each of these components of an atom?
5.2. Describe the difference between an atom and a molecule.
5.3. The number of _____ determines what the element will be.
5.4. The _____ has no electrical charge.
5.5. The _____ has a negative charge.

CHAPTER 6

Thermal Properties and States of Matter

This grand show is eternal. It is always sunrise somewhere; the dew is never all dried at once; a shower is forever falling; vapor is ever rising. Eternal sunrise, eternal sunset, eternal dawn and gloaming, on sea and continents and islands, each in its turn, as the round earth rolls.

—John Muir

Thermal Properties

Thermal properties of chemicals and other substances are important in physics. Such knowledge is used in math calculations, in the study of the physical properties of materials, in hazardous materials spill mitigation, and in solving many other complex environmental problems. *Heat* is a form of energy—thermal energy that can be transferred between two bodies that are at different temperatures. Whenever work is performed, usually a substantial amount of heat is caused by friction. The conservation of energy law tells us the work done plus the heat energy produced must equal the original amount of energy available. That is:

$$\text{total energy} = \text{work done} + \text{heat produced} \qquad (6.1)$$

A traditional unit for measuring heat energy is the *calorie*. A calorie (cal) is defined as the amount of heat necessary to raise the temperature of 1 gram of pure liquid water by 1°C at normal atmospheric pressure. In SI units,

$$1 \text{ cal} = 4.186 \text{ J (joule)}$$

The calorie we have defined should not be confused with the one used when discussing diets and nutrition. A kilocalorie is 1000 calories as we have defined it—the amount of heat necessary to raise the temperature of one kilogram of water by 1°C.

In the British system of units, the unit of heat is the British thermal unit, or *Btu*. One Btu is the amount of heat required to raise the temperature of 1 pound of water 1°F at normal atmospheric pressure (1 atm).

Heat can be transferred in three ways: by conduction, convection, and radiation. When direct contact between two physical objects at different temperatures occurs, heat is transferred via *conduction* from the hotter object to the colder one. When a gas or liquid is placed between two solid objects, heat is transferred by *convection*. Heat is also transferred when no physical medium exists, by *radiation* (for example, radiant energy from the sun).

SPECIFIC HEAT

Earlier we pointed out that one kilocalorie of heat is necessary to raise the temperature of 1 kilogram of water 1°C. Other substances require different amounts of heat to raise the temperature of 1 kilogram of the substance 1°C. The *specific heat* of a substance is the amount of heat in kilocalories necessary to raise the temperature of 1 kilogram of the substance 1°C.

The units of specific heat are Kcal/kg°C or, in SI units, J/kg°C. The specific heat of pure water, for example, is 1.000 kcal/kg°C, or 4186 J/kg°C.

The greater the specific heat of a material, the more heat is required. Also, the greater the mass of the material or the greater the temperature change desired, the more heat is required.

If the specific heat of a substance is known, it is possible to calculate the amount of heat required to raise the temperature of that substance. In general, the amount of heat required to change the temperature of a substance is proportional to the mass of the substance and the change in temperature, according to the following relation:

$$Q = mc \, \Delta T \tag{6.2}$$

where

Q = heat required
m = mass of the substance
c = specific heat of the substance
ΔT = change in temperature

The amount of heat necessary to change one kilogram of a solid into a liquid at the same temperature is called the *latent heat of fusion* of the substance. The temperature of the substance at which this change from solid to liquid takes place is known as the *melting point*. The amount of heat necessary to change one kilogram of a liquid into a gas is called the *latent heat of vaporization*. When this point is reached, the entire mass of substance is in the gas state. The temperature of the substance at which this change from liquid to gas occurs is known as the *boiling point*.

States of Matter

The three common states (or phases) of matter (solid, liquid, gaseous) each have unique characteristics. In the *solid* state, the molecules or atoms are in a relatively fixed position. The molecules are vibrating rapidly, but about a fixed point. Because of this definite position of the molecules, a solid holds its shape. A solid occupies a definite amount of space and has a fixed shape.

When the temperature of a gas is lowered, the molecules of the gas slow down. If the gas is cooled sufficiently, the molecules slow down so much that they lose the energy needed to move rapidly throughout their container. The gas may turn into liquid. Common liquids are water, oil, and gasoline. A *liquid* is a material that occupies a definite amount of space, but which takes the shape of the container.

In some materials, the atoms or molecules have no special arrangement at all. Such materials are called *gases*. Oxygen, carbon dioxide, and nitrogen are common gases. A *gas* is a material that takes the exact volume and shape of its container.

Although the three states of matter discussed above are familiar to most people, the change from one state to another is of primary interest to environmentalists. Two ways that changing matter from one state to another has impact on environmental concerns include water vapor changing from the gaseous state to liquid precipitation, or a spilled liquid chemical changed to a semi-solid substance by the addition of chemicals to aid in the cleanup effort.

THE GAS LAWS

The atmosphere is composed of a mixture of gases, the most abundant of which are nitrogen, oxygen, argon, carbon dioxide, and water vapor (gases and the atmosphere are addressed in greater detail later). The pressure of a gas is the force that the moving gas molecules exert upon a unit area. A common unit of pressure is newtons per square meter, N/m^2, called a pascal (Pa). An important relationship exists among the pressure, volume, and temperature of a gas. This relation is known as the *ideal gas law* and can be stated as

$$\frac{P_1 V_1}{T_1} = \frac{P_2 V_2}{T_2} \qquad (6.3)$$

where P_1, V_1, and T_1 are pressure, volume, and absolute temperature at time 1, and P_2, V_2, and T_2 are pressure, volume, and absolute temperature at time 2. A gas is called perfect, or ideal, when it obeys this law.

A temperature of 0°C (273 K) and a pressure of 1 atmosphere (atm) have been chosen as *standard temperature and pressure (STP)*. At STP the volume of 1 mole of ideal gas is 22.4 L.

LIQUIDS AND SOLUTIONS

The most common solutions are liquids. However, solutions, which are homogenous mixtures, can be solid, gaseous, or liquid. The substance in excess in a solution is called the *solvent*. The substance dissolved is the *solute*. Solutions in which water is the solvent are called *aqueous solutions*. A solution in which the solute is present in only a small amount is called a *dilute solution*. If the solute is present in large amounts, the solution is called a *concentrated solution*. When the maximum amount of solute possible is dissolved in the solvent, the solution is called a *saturated solution*.

The concentration (the amount of solute dissolved) is frequently expressed in terms of the *molar concentration*. The molar concentration, or *molarity*, is the number of moles of solute per liter of solution. Thus, a 1-molar solution, written 1.0 M, has 1 gram formula weight of solute dissolved in 1 liter of solution. In general

$$\text{Molarity} = \frac{\text{moles of solute}}{\text{number of liters of solution}} \qquad (6.4)$$

Note that the number of liters of solution, not the number of liters of solvent, is used.

Example
Problem
Exactly 40 g of sodium chloride (NaCl), or table salt, are dissolved in water and the solution is made up to a volume of 0.80 liter of solution. What is the molar concentration (M) of sodium chloride in the resulting solution?

Solution
First find the number of moles of salt.

$$\text{Number of moles} = \frac{40 \text{ g}}{58.5 \text{ g/mole}} = 0.68 \qquad (6.5)$$

$$\text{Molarity} = \frac{0.68 \text{ mole}}{0.80 \text{ liter}} = 0.85 \text{ M} \qquad (6.6)$$

Problems

6.1. You have 20 grams of water and add 80 calories to it. How much does its temperature rise?
6.2. How many Btu are required to raise the temperature of 10 pounds of water 10°F?
6.3. What does a rating of 14,000 Btu per hour mean on a heater?
6.4. How much heat is required to raise the temperature of 100 grams of iron (specific heat of iron = 0.11) from 30°C to 40°C?
6.5. Why do deserts cool off so fast at night?

CHAPTER 7

Wave Motion and Sound

Now he has departed from this strange world a little ahead of me. That means nothing. People like us, who believe in physics, know that the distinction between past, present, and future is only a stubbornly persistent illusion.

—Albert Einstein

In my industrial hygiene classes at Old Dominion University, before presenting my lecture on the physics of wave motion and sound I ask each student to write down on a piece of paper his or her definition of the term *noise*. The purpose of doing this each semester is to make the point that there is a distinct difference between the two terms *noise* and *sound*. However, it is rare when I actually elicit the definition from my students that I am looking for. That is, defined, noise is any unwanted sound, and this is the point I am trying to make—again, that there is a distinct difference between the two terms.

In this chapter, unlike industrial hygienists and other safety professionals who are concerned with the deleterious effects of noise and noise-induced hearing loss (NIHL), we are concerned with the physics of wave motion and sound.

Waves

Waves and wave motion are basic to an understanding of not only the aspects of sound but also of light, and parts of magnetism and electricity (addressed later). Waves are described as disturbances that travel through a medium from one location to another location. Physical waves (the ones we are concerned with in this text) include water waves, waves on a string, and sound waves. Waves of all types are around us every day. If you are listening to the TV right now, the compressions generated by the motion of the materials in your speakers are sent out in front of the speaker in a pattern that strikes your eardrum; these compressions are called *sound waves*.

WAVE CHARACTERISTICS

Waves have two fundamental characteristics: wavelength and amplitude. The *wavelength* (see figure 7.1) is the length of one complete wave. It depends upon the fre-

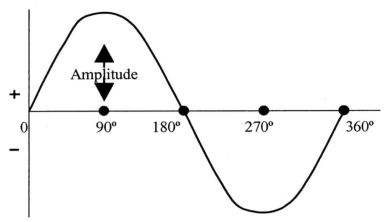

Figure 7.1. Wavelength (0° to 360°) and amplitude.

quency of the periodic variation (period) and its velocity of transmission. The *frequency* (*f*) of the wave measures how many complete cycles of the wave pass a fixed point in one second. The *period* (*T*) of the wave measures the amount of time for the completion of one cycle. The *amplitude* (see figure 7.1) of the wave is the height of the wave, measured as a displacement from some zero level. Wavelength is often abbreviated with the Greek lowercase letter λ (lambda). Expressed as a formula:

$$\lambda = \frac{\text{velocity}}{\text{frequency}} \tag{7.1}$$

Frequency and period are reciprocals of each other, or

$$f = 1/T \tag{7.2}$$

Wave speed can be determined by dividing the wavelength by the period, or

$$v = \lambda/T \tag{7.3}$$

The speed of a wave generally depends on the nature of the medium (the substance that carries a wave from one location to another) through which it is moving. In other words, the speed of sound waves likewise depends on the density and temperature of the air through which waves are traveling.

TRANSVERSE AND LONGITUDINAL WAVES

There are two basic types of physical waves: transverse and longitudinal. A *transverse wave* is a wave in which particles of the medium (in water) move in a direction perpendicular to the direction in which the wave moves.

A *longitudinal wave* is a wave in which particles of the medium move in a direction parallel to the direction in which the wave moves. Sound waves are longitudinal waves. Sound waves arise when the medium is alternately compressed and stretched—like the backward and forward motion of a speaker.

Physics of Sound

Humans are equipped with ears for the detection of sound. Sound is part of our everyday sensory experience. Sound is a wave originating from the vibration of a source that causes pressure changes in air (the medium), which results in pressure waves of varying characteristics. These pressure wave characteristics include amplitude and frequency. In regard to sound, *amplitude* (which determines its volume or intensity on a logarithmic scale) is the amount of sound pressure measured in *decibels* (dB), named in honor of Alexander Graham Bell (1847–1922).

On the logarithmic decibel intensity scale, a sound that is 10 dB is 10 times as loud as a 0-dB sound. A sound that is 20 dB is 10 times louder than a 10-dB sound and 100 times louder than a 0-dB sound. Thus, it follows that on the same scale a loudness of 50 decibels is 100,000 times as loud as a 0-dB sound.

The *frequency* of sound is the rate of vibration per unit time measured in cycles per second, more commonly known as hertz (Hz). The hearing acuity range of a healthy newborn child is 0 Hz to 20,000 Hz (frequencies greater than 20,000 Hz are said to be ultrasonic waves); the range of normal perception for older children is 20 Hz to 20,000 Hz; hearing acuity of older humans is dependent upon normal aging, health issues, and history of noise exposure.

On the decibel scale a faint audible sound measures about 0 dB, a whisper is about 20 dB, and ordinary conversation is about 60 dB. Sound becomes painful at about 120 dB to140 dB, the "threshold of pain." At a distance of a few yards from the stage, a rock band's amplified music has an intensity of about 115 dB to120 dB.

OCTAVE BANDS

Sound is more difficult to measure than pressure or temperature. Sound occurs over a range of distinct frequencies (or *f*); therefore, its level must be measured at each frequency to understand how it will be perceived in a particular environment. Octave bands (standardized notation—characterized by center frequency) quantify effective frequencies without looking at each frequency one at a time.

Why do we need and use octave bands? Simply, as mentioned, humans can sense sounds at frequencies ranging from 20 Hz to 20,000 Hz, but the standard room acoustic-range focuses on the 44 Hz to 11,300 Hz range. Even though this range is less than the range of human hearing (20 Hz – 20,000 Hz) it is still too large to measure—it contains too many data points. In order to make the data more manageable, the 44-Hz to 11,300-Hz spectrum is divided into octave bands. As mentioned, each octave band is characterized by its center frequency and is delimited such that the band's higher frequency is twice its lowest frequency. The "octave band center frequency" is

$$f_c = (f_1 f_2)^{1/2}$$

where

f_c = center frequency
f_1 = lower band edge
f_2 = upper band edge

The center frequencies usually listed include:

Hz									
31.5	62	125	250	500	1K	2K	4K	8K	16K

SOUND PRESSURE

Sound pressure is fundamental to acoustics and is what the human ear hears. It is a pressure disturbance in the atmosphere whose intensity is influenced not only by the strength of the source, but also by the surroundings and the distance from the source to the receiver. Sound pressure is defined as

pressure = force per unit of area

and is in measured in the following units:

• newtons per square meter (N/m^2)—called a pascal (Pa)
• dynes per square centimeter (D/cm^2)

To determine sound pressure level we use

$$20 \, \text{Log} \, (p/p_o)$$

where

p = sound pressure
p_o = reference (equal to the threshold of human hearing, 0.00002 Pa or 20 μPa)

Example 7.1
Problem
If sound pressure is 0.02 Pa, what is the sound pressure level?

Solution

$$20 \times \text{Log} \ (0.02 \ \text{Pa}/0.00002 \ \text{Pa}) \ = \ 60 \ \text{dB}$$

Example 7.2
Problem
If sound pressure is 0.07 Pa, what is the sound pressure level?

Solution

$$20 \times \text{Log} \ (0.07 \ \text{Pa}/0.00002 \ \text{Pa}) \ = \ 71 \ \text{dB (rounded)}$$

Because sound pressure levels (SPLs) are based on a logarithmic scale, they cannot be added directly. For example, 90 dB + 90 dB ≠ 180 dB. To add these SPLs we use the following operation/equation:

$$\text{SPL}_T \ = \ 10 \times \text{Log} \ [\Sigma \ 10 \ (\text{SPL}_i/10)]$$

where

SPL_T = total sound pressure level
SPL_i = the *ith* sound pressure to be summed

Example 7.3
Problem
Given two machines producing 80 dB each, what is the total SPL?

Solution

$$\begin{aligned}
\text{SPL}_T \ &= \ 10 \times \text{Log} \ [\Sigma \ 10 \ (\text{SPL}_i/10)] \\
&= \ 10 \times \text{Log} \ (10^{(80/10)} \ + \ 10^{(80/10)}) \\
&= \ 10 \times \text{Log} \ (2 \times 10^8) \\
&= \ 83 \ \text{dB}
\end{aligned}$$

Example 7.3 points out an important point or rule of thumb: adding two SPLs of equal value will always result in a 3 dB increase!

$$90 \ \text{dB} + 90 \ \text{dB} = 93 \ \text{dB}$$
$$50 \ \text{dB} + 50 \ \text{dB} = 53 \ \text{dB}$$
$$100 \ \text{dB} + 100 \ \text{dB} = 103 \ \text{dB}$$

Example 7.4
Problem
Four machines produce 100 dB, 91 dB, 90 dB, and 89 dB respectively. What is the total sound pressure level?

Solution

$$SPL_T = 10 \times Log [\Sigma 10 (SPL_i)]$$
$$= 10 \times Log [10^{(100/10)} + 10^{(91/10)} + 10^{(90/10)} + 10^{(89/10)}]$$
$$= 10 \times Log [10^{10} + 10^{9.1} + 10^{9} + 10^{8.9}]$$
$$= 101.2 \ dB$$

We end the chapter with a practical point about sound. It should not surprise you that the farther you get from the source of a sound the quieter it becomes. The intensity of a sound wave decreases as the distance squared. This is the case because sound waves move away from a sound source in all directions. The energy of the sound wave decreases with the area through which it passes; that is, the energy from the sound wave is diluted because it passes through increasingly larger areas centered on the sound source. For example, if you were 1 ft from the sound source, the energy passes through a sphere with a 1-ft radius. If you are 3 ft from the sound source, the energy passes through a sphere with a 3-ft radius. Surface area of the sphere is determined by:

$$4\pi r^2$$

The intensity decreases with the square of the distance from the sound source.

Problems

7.1. If wave crests in a pond are 3 m apart and the period of the waves is 1 s, what is the wave velocity?
7.2. Does temperature have an effect on the speed of sound in a medium? Why?
7.3. Are sound waves transverse or longitudinal?
7.4. Sound intensity increases or decreases with sound _____.
7.5. A 50-dB sound carries how many times as much energy as a 0-dB sound?

CHAPTER 8

Light and Color

> There are children playing in the street who could solve some of my top problems in physics, because they have modes of sensory perception that I lost long ago.
>
> —Julius Robert Oppenheimer

The rays of light from the sun are reflected from all the surfaces that we see, and these reflected rays enter our eyes. Through refraction the lenses in our eyes focus the rays of light onto our retinas, where the energy of the light is deposited and converted by our brains into both intensity and color. At nighttime our sight is predicated on the availability of artificial light from candles, streetlights, flashlights, lightbulbs, or some other source(s). Simply, if we do not have some sort of light to reflect off surfaces and enter the lenses of our eyes, then we have nothing but darkness.

In this chapter, we examine the properties of light. We study light as streams of particles that move in straight lines, reflecting off surfaces or bending through materials, as they travel. We finish our discussion of light with color and the wave properties of light.

Light

Light is all around us, but it is difficult to define and to explain. For example, light is electromagnetic radiation that can be detected by the human eye. But then what is electromagnetic radiation? Electromagnetic radiation is both magnetic and electric fields that change and propagate through space. This radiation is an electromagnetic wave that has amplitude (the brightness of the light) and wavelength (the color of the light). Other aspects about light are more obvious, however. It is obvious to us that light emanates from some source such as the sun or a lightbulb. The shadow of our body on the sidewalk or trail on a sunny day is evidence that light rays originated from the sun. When we see our shadow on some surface at night, it is evidence that other sources besides the sun can produce light. We know that an object is in front of or near us because light carries information about the object from which it is reflected. We also know that light is changed as light rays bounce off an object or travel through a substance. Through observation and research we know (have learned) that light has the properties of both waves and particles.

SPEED OF LIGHT

The speed of light, usually denoted by the letter c, in a vacuum is 183,310 mi/s or 299,776 km/s. Light that travels through regions other than a vacuum is slower. For example, in water, light is slowed by 25 percent; in glass, by 35 percent; and in air, by just 0.03 percent.

SOURCES OF LIGHT

The primary source of light on Earth is the sun. Sunlight is the most common type of light that we see every day. Sunlight, regular lightbulbs (not fluorescent), and fires are all incandescent sources of light. Incandescent light sources produce light through vibration of entire atoms.

Another type of light source is luminescence, which does not rely on the vibration of entire atoms but instead involves only the electrons. Luminescence requires continuous excitation to give electrons a boost to a higher energy level to keep producing light. This boost may be provided by many sources: electrical current as in neon light, fluorescent lights, mercury-vapor streetlights, television screens, light-emitting diodes (LEDs), and computer monitors; chemical reactions produced by fireflies; or radioactivity as in luminous paints, to name just a few.

LUMINOUS INTENSITY OF LIGHT

The intrinsic brightness or *luminous intensity* (L_v), of a light source is quantified by a measure of the wavelength-weight power emitted by a light source in a particular direction per unit solid angle. The SI base unit of luminous intensity is the standard candle (dates to a time when a wax candle was the standard source of light) or *candela*, abbreviated *cd*. One candle or candela is equal to 1/60 of the luminous intensity per square centimeter of a perfect emitter of radiation (aka *black-body*) at a specific temperature. *Luminous flux (F)* (also called *luminous power*) is the rate of flow (in *lumens*) of energy carried by light

✔ **Interesting Point:** The intensity of a lamp can be measured by comparing the illumination it produces with that from a standard lamp by means of a device called a *photometer*.

The relationship among illumination, luminous intensity, and distance is given by:

$$E = C/d^2 \tag{8.1}$$

where

E = illumination (foot-candles)
C = luminous intensity (candles)
d = distance (feet)

Example 8.1
Problem
How far away does a lamp with a 100-cd bulb need to be placed to provide 25 ft-c (candles per square foot) of illumination?

Solution
Use the relationship $E = C/d^2$.

$$25 \text{ cd/ft}^2 = 100 \text{ cd}/d^2$$

or

$$d^2 = 100 \text{ cd/ft}^2/25 \text{ cd} = 4 \text{ ft}^2$$
$$d = 2 \text{ ft}$$

Therefore, the lamp needs to be placed at a distance of 2 ft in order to have a 25 ft-c level of illumination.

LAWS OF REFLECTION AND REFRACTION

In order to better understand the information presented in this section it is important to define key terms specific to the properties of reflection and refraction of light.

Key Terms

- *Reflection*: the process by which radiation that strikes a surface separating two media (e.g., air and glass) of different densities is in part or in whole turned back into the medium from which it originated. The radiation rebounds from a barrier in its path without a change in speed.
- *Refraction*: a change in the direction of rays passing through a boundary at an angle to the normal.
- *Normal line*: a line perpendicular to the surface at the point where the radiation strikes (see figures 8.1 and 8.2).
- *Ray*: an idealized, straight, narrow beam of light.
- *Incident ray*: a ray of light that strikes (impinges upon) a surface. The angle between this ray and the perpendicular or normal to the surface is the angle of incidence.
- *Reflected ray*: a ray that has rebounded from a surface (see figures 8.1 and 8.2).
- *Angle of incidence*: the angle between the incident ray and a normal line (see figures 8.1 and 8.2).
- *Angle of reflection*: the angle between the reflected ray and the normal line (see figure 8.1).
- *Angle of refraction*: the angle between the refracted ray and the normal line (see figure 8.2).
- *Index of refraction* : the ratio speed of light (c) in a vacuum to its speed (v) in a given material; always greater than 1.

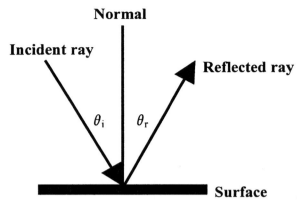

Figure 8.1. Law of reflection.

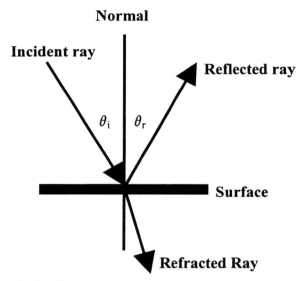

Figure 8.2. Law of refraction.

Laws of Reflection

Objects can be seen by the light they emit, or, more often, by the light they reflect. As mentioned, *reflection* is the change in the direction of rays that bounce off a boundary between two substances. Reflected light obeys the law of reflection, that the angle of reflection equals the angle of incidence. The laws of reflection state:

• As shown in figure 8.1, the incident ray, normal line, and reflected ray are in the same plane.

- The angle of incidence and angle of reflection are equal in magnitude and in opposite directions from the normal line (see figure 8.1).
- If there is no change in the transmitting, the reflected ray travels with the same speed as the incident ray ($\theta_i = \theta_r$).

Laws of Refraction

Simply, the laws of reflection state that for a medium in which the value of a particular property is independent of the direction of measurement:

- A ray of light that passes through a surface into a denser medium (e.g., from air to glass) is refracted toward the normal line. A ray that passes through a surface to a less dense medium (e.g., from glass to air), is refracted away from the normal line.
- The incident and refracted rays are in the same plane (see figure 8.2).
- The ratio of the sine of the angle of incidence to the sine of the angle of refraction is a constant, called the relative index of refraction, equal to the ratio of the indices of refraction for the two media.

POLARIZATION

When light waves (which can vibrate in many directions) vibrate in one direction (in a single plane such as up and down), they are called *polarized light*. Light waves that vibrate in more than one direction (in more than one plane, such as both up/down and right/left—vibrating in a vertical and a horizontal plane) are called *unpolarized light*.

To polarize light, the light must be passed through a polaroid filter that will only allow light of single polarity to pass. The resulting light will be polarized light of half intensity. When two polaroid filters are used and positioned so that one is rotated 90° to the other, no light will be able to pass.

Color

The sun generates light that we typically assume is white light. The reality is that sunlight actually contains the entire spectrum of light (visible light) or colors: red, orange, yellow, green, blue, indigo, and violet. During grade school or high school earth science classes when studying light, color, and the color spectrum, many students learn and memorize the mnemonic (consisting of the first letter in the name of each color) for the sequence of hues in the visible spectrum, in rainbows, and in order from longest to shortest wavelength:

ROY G. BIV = Red, Orange, Yellow, Green, Blue, Indigo, Violet

In terms of visible light, color is simply electromagnetic energy of s specific wave-length. Within the visible part of the spectrum (between 400 and 700 nm), the color perceived by the eye is dependent on the wavelength of the light. Thus, objects that give off (or reflect) green light look green. Objects that give off (or reflect) violet light look violet.

✔ **Important Point:** The visible spectrum actually consists of an infinite gradation of color from red to violet—the ROY G. BIV divisions are for convenience only.

Problems

8.1. Assume that two light towers are separated by 5 mi. How long will it take for the light released when one of the tower lights is energized to reach the other one?

8.2. What is the luminous intensity (in candles) required to produce 20 ft-c of illumination 40 ft below a street lamp?

8.3. If the wavelength of a light is 656 nm, what is the frequency of the light?

8.4. What is the frequency of radio waves, assuming that their wavelength is 20 cm?

8.5. The quantity that specifies the intrinsic brightness of a light source is _____.

CHAPTER 9

Electricity

Why, sir, there is every possibility that you will soon be able to tax it!
(to Prime Minister William Gladstone, on the usefulness of electricity)

—Michael Faraday (1791–1867)

Electricity: What Is It?

Electricity, electric charge, electric currents, electric power, batteries, and electromagnetic signals are so much a part of our daily lives that we barely notice them anymore—unless our electrical devices fail to work, of course. We are literally surrounded by electrical phenomena in both the natural world and human-made world. Although, as just stated, the wonders of electricity are often ignored and taken for granted, when electrical machines, equipment, and/or appliances actually do get someone's attention, those paying attention, living, and/or working in modern societies generally have little difficulty in recognizing electrical equipment.

Most of us know what electricity does for us. Consider the typical industrial worker in a typical industrial workplace, for example. Industrial plants are equipped with electrical equipment that

- generates electricity (a generator or emergency generator)
- stores electricity (batteries)
- changes electricity from one level (voltage or current) to another (transformers)
- transports or transmits and distributes electricity throughout the plant site (wiring distribution systems)
- measures electricity (meters)
- converts electricity into other forms of energy (rotating shafts—mechanical energy, heat energy, light energy, chemical energy, or radio energy)
- protects other electrical equipment (fuses, circuit breakers, or relays)
- operates and controls other electrical equipment (motor controllers)
- converts some condition or occurrence into an electric signal (sensors)
- converts some measured variable to a representative electrical signal (transducers or transmitters)

Recognizing electrical equipment is easy because we use so much of it. If we ask typical industrial workers where such equipment is located in their plant site, they know, because they either operate or perform maintenance on these same devices or their ancillaries. If we ask these same workers what a particular electrical device does, they can tell us. If we were to ask if their plant electrical equipment was important to plant operations, the chorus would resound, "Absolutely."

Here's another question that does not always result in such a resounding note of assurance, however. If we were to ask these same workers to explain to us in very basic terms how electricity works to make their plant equipment operate, the answers we probably would receive would be varied, jumbled, disjointed—and probably not all that accurate. Even on a more basic level, how many workers would be able to accurately answer the question "What is electricity?"

NATURE OF ELECTRICITY

The word *electricity* is derived from the Greek word "electron" (meaning "amber"). Amber is a translucent (semitransparent), yellowish, fossilized mineral resin. The ancient Greeks used the words "electric force" in referring to the mysterious forces of attraction and repulsion exhibited by amber when it was rubbed with a cloth. They did not understand the nature of this force. They could not answer the question "What is electricity?" This question still remains unanswered. Today, we often attempt to answer this question by describing the effect and not the force. That is, the standard answer given in physics is: "The force that moves electrons" is electricity, which is about the same as defining a sail as "The force that moves a sailboat."

At the present time, little more is known than the ancient Greeks knew about the fundamental nature of electricity, but we have made tremendous strides in harnessing and using it. As with many other unknown (or unexplainable) phenomena, elaborate theories concerning the nature and behavior of electricity have been advanced, and have gained wide acceptance because of their apparent truth—and because they work.

Scientists have determined that electricity seems to behave in a constant and predictable manner in given situations or when subjected to given conditions. Faraday, Ohm, Lenz, and Kirchhoff have described the predictable characteristics of electricity and electric current in the form of certain rules. These rules are often referred to as laws. Thus, though electricity itself has never been clearly defined, its predictable nature and easily used form of energy has made it one of the most widely used power sources in modern times.

The bottom line: You can "learn" about electricity by learning the rules, or laws, applying to the behavior of electricity and by understanding the methods of producing, controlling, and using it. Thus, this learning can be accomplished without ever having determined its fundamental identity.

You are probably scratching your head.

We understand the main question running through your human microprocessor at this exact moment: "This is a chapter about the physics of electricity and the author can't even explain what electricity is?"

That is correct; I cannot. The point is that no one can definitively define electricity. Electricity is one of those subject areas where the old saying "We don't know what we don't know about it" fits perfectly.

Again, there are a few theories about electricity that have so far stood the test of extensive analysis and much time (relatively speaking, of course). One of the oldest and the most generally accepted theory concerning electric current flow (or electricity) is known as the *electron theory*.

The electron theory basically states that electricity or current flow is the result of the flow of free electrons in a conductor. Thus, electricity is the flow of free electrons or simply electron flow. And, in this text, this is how we define electricity—that is, again, electricity is the flow of free electrons.

Electrons are extremely tiny particles of matter. To gain understanding of electrons and exactly what is meant by "electron flow," it is necessary to briefly review our earlier discussion about the structure of matter.

THE STRUCTURE OF MATTER

Matter is anything that has mass and occupies space. To study the fundamental structure or composition of any type of matter, it must be reduced to its fundamental components. All matter is made of *molecules*, or combinations of *atoms* (Greek: "not able to be divided") that are bound together to produce a given substance, such as salt, glass, or water. For example, if you keep dividing water into smaller and smaller drops, you would eventually arrive at the smallest particle that was still water. That particle is the molecule, which is defined as "the smallest bit of a substance that retains the characteristics of that substance."

A molecule of water (H_2O) is composed of one atom of oxygen and two atoms of hydrogen. If the molecule of water were further subdivided, there would remain only unrelated atoms of oxygen and hydrogen, and the water would no longer exist as such. Thus, the molecule is the smallest particle to which a substance can be reduced and still be called by the same name. This applies to all substances—solids, liquids, and gases.

✔ **Important Point:** Molecules are made up of atoms, which are bound together to produce a given substance.

Recall that atoms are composed, in various combinations, of subatomic particles of *electrons, protons,* and *neutrons.* These particles differ in weight (a proton is much heavier than the electron) and charge. We are not concerned with the weights of particles in this text, but the *charge* is extremely important in electricity. The electron is the fundamental negative charge ($-$) of electricity. Electrons revolve about the nucleus or center of the atom in paths of concentric *orbits*, or shells. The proton is the fundamental positive ($+$) charge of electricity. Protons are found in the nucleus. The number of protons within the nucleus of any particular atom specifies the atomic number of that atom. For example, the helium atom has 2 protons in its nucleus so the atomic number is 2. The neutron, which is the fundamental neutral charge of electricity, is also found in the nucleus.

Most of the weight of the atom is in the protons and neutrons of the nucleus.

Whirling around the nucleus is one or more negatively charged electrons. Normally, there is one proton for each electron in the entire atom, so that the net positive charge of the nucleus is balanced by the net negative charge of the electrons rotating around the nucleus (see figure 9.1).

✔ **Important Point:** Most batteries are marked with the symbols + and − or even with the abbreviations POS (positive) and NEG (negative). The concept of a positive or negative polarity and its importance in electricity will become clear later. However, for the moment, you need to remember that an electron has a negative charge and that a proton has a positive charge.

We stated earlier that in an atom the number of protons is usually the same as the number of electrons. This is an important point because this relationship determines the kind of element in question (the atom is the smallest particle that makes up an element; an element retains its characteristics when subdivided into atoms). Figure 9.2 shows a simplified drawing of several atoms of different materials based on the conception of electrons orbiting about the nucleus. For example, hydrogen has a nucleus consisting of 1 proton, around which orbits 1 electron. The helium atom has a nucleus containing 2 protons and 2 neutrons with 2 electrons encircling the nucleus. Both of these elements are electrically neutral (or balanced) because each has an equal number of electrons and protons. Since the negative (−) charge of each electron is equal in magnitude to the positive (+) charge of each proton, the two opposite charges cancel.

A balanced (neutral or stable) atom has a certain amount of energy, which is equal to the sum of the energies of its electrons. Electrons, in turn, have different energies called *energy levels*. The energy level of an electron is proportional to its distance from the nucleus. Therefore, the energy levels of electrons in shells farther from the nucleus are higher than that of electrons in shells nearer the nucleus.

When an electric force is applied to a conducting medium, such as copper wire, electrons in the outer orbits of the copper atoms are forced out of orbit (i.e., liberating or freeing electrons) and impelled along the wire. This electrical force, which forces electrons out of orbit, can be produced in a number of ways, such as by moving a

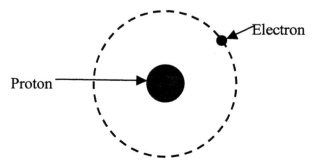

Figure 9.1. One proton and one electron = electrically neutral.

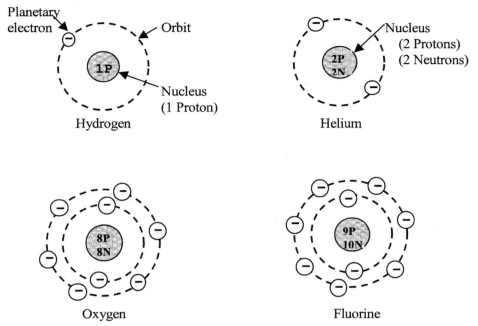

Figure 9.2.   Atomic structure of elements.

conductor through a magnetic field; by friction, as when a glass rod is rubbed with cloth (silk); or by chemical action, as in a battery. When the electrons are forced from their orbits they are called *free electrons*. Some of the electrons of certain metallic atoms are so loosely bound to the nucleus that they are relatively free to move from atom to atom. These free electrons constitute the flow of an electric current in electrical conductors.

✔ **Important Point:** When an electric force is applied to a copper wire, free electrons are displaced from the copper atoms and move along the wire, producing electric current as shown in figure 9.3.

If the internal energy of an atom is raised above its normal state, the atom is said to be *excited*. Excitation may be produced by causing the atoms to collide with particles that are impelled by an electric force as shown in figure 9.3. In effect what occurs

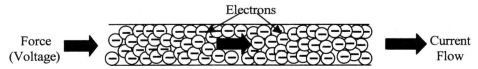

Figure 9.3.   Electron flow in a copper wire.

is that energy is transferred from the electric source to the atom. The excess energy absorbed by an atom may become sufficient to cause loosely bound outer electrons (as shown in figure 9.3) to leave the atom against the force that acts to hold them within.

✔ **Important Point:** An atom that has lost or gained one or more electrons is said to be *ionized.* If the atom loses electrons it becomes positively charged and is referred to as a *positive ion.* Conversely, if the atom gains electrons, it becomes negatively charged and is referred to as a *negative ion.*

STATIC ELECTRICITY

Electricity at rest is often referred to as *static electricity.* More specifically, when two bodies of matter have unequal charges and are near one another, an electric force is exerted between them because of their unequal charges. However, since they are not in contact, their charges cannot equalize. The existence of such an electric force, where current can't flow, is *static electricity.*

However, static, or electricity at rest, will flow if given the opportunity. An example of this phenomenon is often experienced when one walks across a dry carpet and then touches a doorknob—a slight shock is usually felt and a spark at the fingertips is likely noticed. In the workplace, static electricity is prevented from building up by properly bonding equipment to ground or earth.

CHARGED BODIES

To fully grasp a clear understanding of static electricity, it is necessary to know one of the fundamental laws of electricity and its significance.

The fundamental law of charged bodies states: "Like charges repel each other and unlike charges attract each other."

A positive charge and negative charge, being opposite or unlike, tend to move toward each other—attracting each other. In contrast, like bodies tend to repel each other. Electrons repel each other because of their like negative charges, and protons repel each other because of their like positive charges. Figure 9.4 demonstrates the law of charged bodies.

It is important to point out another significant part of the fundamental law of charged bodies—that is, the force of attraction or repulsion existing between two magnetic poles decreases rapidly as the poles are separated from each other. More specifically, the force of attraction or repulsion varies directly as the product of the separate pole strengths and inversely as the square of the distance separating the magnetic poles, provided the poles are small enough to be considered as points. Let's look at an example:

If you increase the distance between two north poles from 2 ft to 4 ft, the force of repulsion between them is decreased to one-fourth of its original value. If either

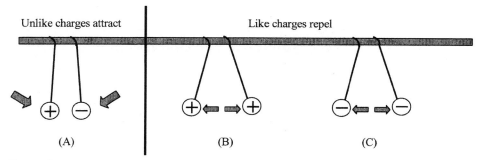

Figure 9.4.   Reaction between two charged bodies. The opposite charge in (A) attracts. The like charges in (B) and (C) repel each other.

pole strength is doubled, the distance remaining the same, the force between the poles will be doubled.

Coulomb's Law

Simply put, Coulomb's law points out that the amount of attracting or repelling force which acts between two electrically charged bodies in free space depends on two things: their charges and the distance between them.

Specifically, Coulomb's law states: "Charged bodies attract or repel each other with a force that is directly proportional to the product of their charges, and inversely proportional to the square of the distance between them."

✔ **Note:** The magnitude of electric charge a body possesses is determined by the number of electrons compared with the number of protons within the body. The symbol for the magnitude of electric charge is Q, expressed in units of coulombs (C). A charge of one positive coulomb means a body contains a charge of 6.25×10^{18}. A charge of one negative coulomb, $-Q$, means a body contains a charge of 6.25×10^{18} more electrons than protons.

Electrostatic Fields

The fundamental characteristic of an electric charge is its ability to exert force. The space between and around charged bodies in which their influence is felt is called an *electric field of force*. The electric field is always terminated on material objects and extends between positive and negative charges. This region of force can consist of air, glass, paper, or a vacuum. This region of force is referred to as an *electrostatic field*.

When two objects of opposite polarity are brought near each other, the electrostatic field is concentrated in the area between them. The field is generally represented by lines that are referred to as *electrostatic lines of force*. These lines are imaginary and are used merely to represent the direction and strength of the field. To avoid confusion, the positive lines of force are always shown leaving charge, and for a negative charge they are shown as entering. Figure 9.5 illustrates the use of lines to represent the field about charged bodies.

✔ **Note:** A charged object will retain its charge temporarily if there is no immediate transfer of electrons to or from it. In this condition, the charge is said to be at rest. Remember, electricity at rest is called *static electricity.*

MAGNETISM

Most electrical equipment depends directly or indirectly upon magnetism. Magnetism is defined as a phenomenon associated with magnetic fields; that is, it has the power to attract such substances as iron, steel, nickel, or cobalt (metals that are known as magnetic materials). Correspondingly, a substance is said to be a magnet if it has the property of magnetism. For example, a piece of iron can be magnetized and is thus a magnet.

When magnetized, the piece of iron (we will assume a piece of flat bar 6 inches long × 1 inch wide × 0.5 inches thick—a bar magnet; see figure 9.6) will have two points opposite each other that most readily attract other pieces of iron. The points of maximum attraction (one on each end) are called the *magnetic poles* of the magnet: the north (N) pole and the south (S) pole. Just as like electric charges repel each other and opposite charges attract each other, like magnetic poles repel each other and unlike poles attract each other. Although invisible to the naked eye, magnetic force can be shown to exist by sprinkling small iron filings on a glass covering a bar magnet, as shown in figure 9.6.

Figure 9.7 shows how the field looks without iron filings; it is shown as lines of force (known as magnetic flux or flux lines) in the field, repelled away from the north

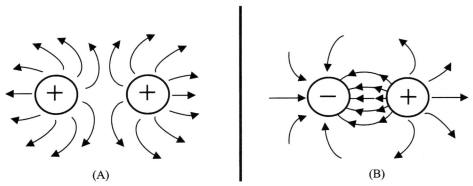

(A)   (B)

Figure 9.5.   Electrostatic lines of force. (A) represents the repulsion of like-charged bodies and their associated fields. (B) represents the attraction between unlike-charged bodies and their associated fields.

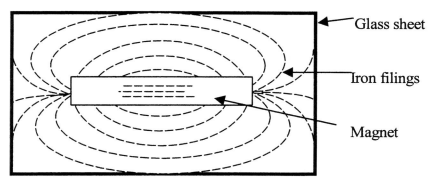

Figure 9.6.   The magnetic field around a bar magnet. If the glass sheet is tapped gently, the filings will move into a definite pattern that describes the field of force around the magnet.

pole of the magnet and attracted to its south pole. The symbol for magnetic flux is the Greek lowercase letter φ (phi).

✒ **Note:** A *magnetic circuit* is a complete path through which magnetic lines of force may be established under the influence of a magnetizing force. Most magnetic circuits are composed largely of magnetic materials in order to contain the magnetic flux. These circuits are similar to the *electric circuit* (an important point), which is a complete path through which current is caused to flow under the influence of an electromotive force.

There are three types or groups of magnets:

1. *Natural magnets*: found in the natural state in the form of a mineral (an iron compound) called magnetite.
2. *Permanent magnets* (artificial magnet): hardened steel or some alloy such as Alinco bars that have been permanently magnetized. The permanent magnet most people are familiar with is the horseshoe magnet (see figure 9.8).
3. *Electromagnets* (artificial magnet): composed of soft-iron cores around which are wound coils of insulated wire. When an electric current flows through the coil, the

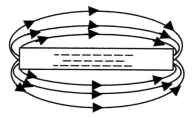

Figure 9.7.   Magnetic field of force around a bar magnet, indicated by lines of force.

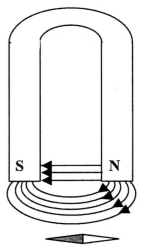

Figure 9.8.   Horseshoe magnet.

core becomes magnetized. When the current ceases to flow, the core loses most of the magnetism.

Magnetic Materials

Natural magnets are no longer used (they have no practical value) in electrical circuitry because more powerful and more conveniently shaped permanent magnets can be produced artificially. Commercial magnets are made from special steels and alloys—magnetic materials.

Magnetic materials are those materials that are attracted or repelled by a magnet and that can be magnetized themselves. Iron, steel, and alloy bar are the most common magnetic materials. These materials can be magnetized by inserting the material (in bar form) into a coil of insulated wire and passing a heavy direct current through the coil. The same material may also be magnetized if it is stroked with a bar magnet. It will then have the same magnetic property that the magnet used to induce the magnetism has—namely, there will be two poles of attraction, one at either end. This process produces a permanent magnet by induction—that is, the magnetism is induced in the bar by the influence of the stroking magnet.

✔ **Note:** Permanent magnets are those of hard magnetic materials (hard steel or alloys) that retain their magnetism when the magnetizing field is removed. A temporary magnet is one that has no ability to retain a magnetized state when the magnetizing field is removed.

Even though they are classified as permanent magnets, it is important to point out that hardened steel and certain alloys are relatively difficult to magnetize and are said to have a *low permeability* because the magnetic lines of force do not easily permeate, or distribute themselves readily through, the steel.

🖝 **Note:** *Permeability* refers to the ability of a magnetic material to concentrate magnetic flux. Any material that is easily magnetized has high permeability. A measure of permeability for different materials in comparison with air or vacuum is called *relative* permeability, symbolized by μ (the Greek letter mu).

Once hard steel and other alloys are magnetized, however, they retain a large part of their magnetic strength and are called *permanent magnets*. Conversely, materials that are relatively easy to magnetize—such as soft iron and annealed silicon steel—are said to have a *high permeability*. Such materials retain only a small part of their magnetism after the magnetizing force is removed and are called *temporary magnets*.

The magnetism that remains in a temporary magnet after the magnetizing force is removed is called *residual magnetism*.

Early magnetic studies classified magnetic materials merely as being magnetic and nonmagnetic—that is, based on the strong magnetic properties of iron. However, since weak magnetic materials can be important in some applications, present studies classify materials into one of three groups: paramagnetic, diamagnetic, and ferromagnetic.

1. *Paramagnetic materials*: These include aluminum, platinum, manganese, and chromium—materials that become only slightly magnetized even though under the influence of a strong magnetic field. This slight magnetization is in the same direction as the magnetizing field. Relative permeability is slightly more than 1 (i.e., considered nonmagnetic materials).
2. *Diamagnetic materials*: These include bismuth, antimony, copper, zinc, mercury, gold, and silver—materials that can also be slightly magnetized when under the influence of a very strong field. Relative permeability is less than 1 (i.e., considered nonmagnetic materials).
3. *Ferromagnetic materials*: These include iron, steel, nickel, cobalt, and commercial alloys—materials that are the most important group for applications of electricity and electronics. Ferromagnetic materials are easy to magnetize and have high permeability, ranging from 50 to 3000.

Magnetic Earth

The earth is a huge magnet, and surrounding the earth is the magnetic field produced by the earth's magnetism. Most people would have no problem understanding or at least accepting this statement. However, if told that the earth's north magnetic pole is actually its south magnetic pole and that the south magnetic pole is actually the earth's north magnetic pole, they might not accept or understand this statement. But, in terms of a magnet, it is true.

As can be seen from figure 9.9, the magnetic polarities of the earth are indicated. The geographic poles are also shown at each end of the axis of rotation of the earth. Clearly, as shown in figure 9.9, the magnetic axis does not coincide with the geographic axis, and therefore the magnetic and geographic poles are not at the same place on the surface of the earth.

Recall that magnetic lines of force are assumed to emanate from the north pole of a magnet and to enter the south pole as closed loops. Because the earth is a magnet, lines of force emanate from its north magnetic pole and enter the south magnetic pole as closed loops. A compass needle aligns itself in such a way that the earth's lines of force enter at its south pole and leave at its north pole. Because the north pole of the needle is defined as the end that points in a northerly direction it follows that the magnetic pole in the vicinity of the north geographic pole is in reality a south magnetic pole, and vice versa.

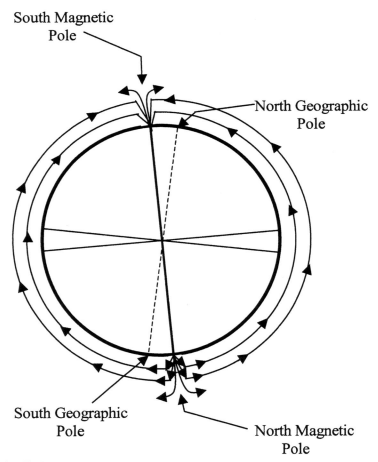

Figure 9.9. Earth's magnetic poles.

DIFFERENCE IN POTENTIAL

Because of the force of its electrostatic field, an electric charge has the ability to do the work of moving another charge by attraction or repulsion. The force that causes free electrons to move in a conductor as an electric current may be referred to as follows:

- Electromotive force (emf)
- Voltage
- Difference in potential

When a difference in potential exists between two charged bodies that are connected by a wire (conductor), electrons (current) will flow along the conductor. This flow is from the negatively charged body to the positively charged body, until the two charges are equalized and the potential difference no longer exists.

✔ **Note:** The basic unit of potential difference is the *volt* (V). The symbol for potential difference is V, indicating the ability to do the work of forcing electrons (current flow) to move. Because the volt unit is used, potential difference is called *voltage*.

The Water Analogy

In attempting to train individuals in the concepts of basic electricity, especially in regard to difference of potential (voltage), current, and resistance relationships in a simple electrical circuit, it has been common practice to use what is referred to as the water analogy. We use the water analogy later to explain (in simple straightforward fashion) voltage, current, and resistance and their relationships in more detail, but for now we use the analogy to explain the basic concept of electricity: difference of potential, or voltage. Because a difference in potential causes current flow (against resistance), it is important that this concept be understood first before the concept of current flow and resistance are explained.

Consider the water tanks shown in figure 9.10—two water tanks connected by a pipe and valve. At first the valve is closed and all the water is in Tank A. Thus, the water pressure across the valve is at maximum. When the valve is opened, the water flows through the pipe from A to B until the water level becomes the same in both tanks. The water then stops flowing in the pipe, because there is no longer a difference in water pressure (difference in potential) between the two tanks.

Just as the flow of water through the pipe in figure 9.10 is directly proportional to the difference in water level in the two tanks, current flow through an electric circuit is directly proportional to the difference in potential across the circuit.

✔ **Important Point:** A fundamental law of current electricity is that the current is directly proportional to the applied voltage—that is, if the voltage is increased, the current is increased. If the voltage is decreased, the current is decreased.

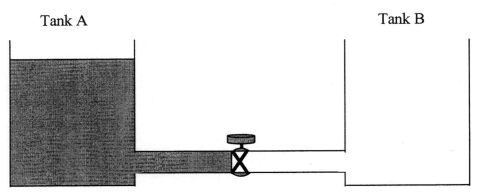

Figure 9.10.   Water analogy of electric difference of potential.

Principal Methods of Producing a Voltage

There are many ways to produce electromotive force, or voltage. Some of these methods are much more widely used than others. The following is a list of the six most common methods of producing electromotive force:

1. *Friction*: voltage produced by rubbing two materials together.
2. *Pressure (piezoelectricity)*: voltage produced by squeezing crystals of certain substances.
3. *Heat (thermoelectricity)*: voltage produced by heating the joint (junction) where two unlike metals are joined.
4. *Light (photoelectricity)*: voltage produced by light striking photosensitive (light-sensitive) substances.
5. *Chemical action*: voltage produced by chemical reaction in a battery cell.
6. *Magnetism*: voltage produced in a conductor when the conductor moves through a magnetic field, or a magnetic field moves through the conductor in such a manner as to cut the magnetic lines of force of the field.

In the study of the physics of basic electricity, we are most concerned with magnetism (generators) and chemistry (batteries) as means to produce voltage. Friction has little practical application, though we discussed it earlier in studying static electricity. Pressure, heat, and light do have useful applications, but we do not need to consider them in this text. Magnetism and chemistry, on the other hand, are the principal sources of voltage and are discussed at length in this text.

ELECTRIC CURRENT

The movement or the flow of electrons is called *current*. To produce current, the electrons must be moved by a potential difference.

✔ **Note:** The terms *current, current flow, electron flow, electron current*, and so on, may be used to describe the same phenomenon.

Electron flow, or current, in an electric circuit is from a region of less negative potential to a region of more positive potential.

✔ **Note:** Current is represented by the letter I. The basic unit in which current is measured is the *ampere*, or *amp* (A). One ampere of current is defined as the movement of one coulomb past any point of a conductor during one second of time.

Recall that we used the water analogy to help us understand potential difference. We can also use the water analogy to help us understand current flow through a simple electric circuit.

Consider figure 9.11, which shows a water tank connected via a pipe to a pump with a discharge pipe. If the water tank contains an amount of water above the level of the pipe opening to the pump, the water exerts pressure (a difference in potential) against the pump. When sufficient water is available for pumping with the pump, water flows through the pipe against the resistance of the pump and pipe. The analogy should be clear: in an electric circuit if a difference of potential exists, current will flow in the circuit.

Another simple way of looking at this analogy is to consider figure 9.12, where the water tank has been replaced with a generator, the pipe with a conductor (wire), and water flow with the flow of electric current.

Again, the key point illustrated by figures 9.11 and 9.12 is that to produce current, the electrons must be moved by a potential difference.

Electric current is generally classified into two general types: direct current (d.c.) and alternating current (a.c.). *Direct current* is current that moves through a conductor or circuit in one direction only. *Alternating current* periodically reverses direction.

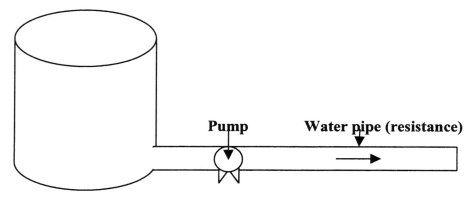

Figure 9.11. Water analogy: current flow.

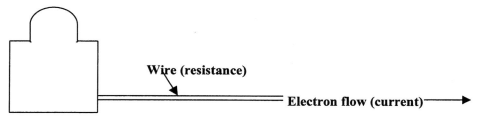

Figure 9.12.   Simple electric circuit with current flow.

RESISTANCE

Earlier it was pointed out that free electrons, or electric current, could move easily through a good conductor, such as copper, but that an insulator, such as glass, was an obstacle to current flow. In the water analogy shown in figure 9.11 and the simple electric circuit shown in figure 9.12, resistance is indicated by either the pipe or the conductor.

Every material offers some resistance, or opposition, to the flow of electric current through it. Good conductors, such as copper, silver, and aluminum, offer very little resistance. Poor conductors, or insulators, such as glass, wood, and paper, offer a high resistance to current flow.

✔ Note: The amount of current that flows in a given circuit depends on two factors: voltage and resistance.

✔ Note: Resistance is represented by the letter R. The basic unit in which resistance is measured is the *ohm* (Ω). One ohm is the resistance of a circuit element, or circuit, which permits a steady current of 1 ampere (1 coulomb per second) to flow when a steady electromotive force (emf) of 1 volt is applied to the circuit. Manufactured circuit parts containing definite amounts of resistance are called *resistors*.

The size and type of material of the wires in an electric circuit are chosen so as to keep the electrical resistance as low as possible. In this way, current can flow easily through the conductors, just as water flows through the pipe between the tanks in figure 9.10. If the water pressure remains constant, the flow of water in the pipe will depend on how far the valve is opened. The smaller the opening, the greater the opposition (resistance) to the flow, and the smaller will be the rate of flow in gallons per second.

In the electric circuit shown in figure 9.12, the larger the diameter of the wire, the lower will be its electrical resistance (opposition) to the flow of current through it. In the water analogy, pipe friction opposes the flow of water between the tanks. This friction is similar to electrical resistance. The resistance of the pipe to the flow of water through it depends upon (1) the length of the pipe, (2) the diameter of the pipe, and (3) the nature of the inside walls (rough or smooth). Similarly, the electrical resistance of the conductors depends upon (1) the length of the wires, (2) the diameter of the wires, and (3) the material of the wires (copper, silver, etc.).

It is important to note that temperature also affects the resistance of electrical conductors to some extent. In most conductors (copper, aluminum, etc.), the resistance increases with temperature. Carbon is an exception. In carbon the resistance decreases as temperature increases.

✔ **Important Note:** Electricity is a study that is frequently explained in terms of opposites. The term that is exactly the opposite of resistance is *conductance*. Conductance (G) is the ability of a material to pass electrons. The unit of conductance is the *Mho*, which is *ohm* spelled backwards. The relationship that exists between resistance and conductance is the reciprocal. A reciprocal of a number is obtained by dividing the number into 1. If the resistance of a material is known, dividing its value into 1 will give its conductance. Similarly, if the conductance is known, dividing its value into 1 will give its resistance.

Battery-Supplied Electricity

Battery-supplied direct current (d.c.) electricity has many applications and is widely used. Applications include providing electrical energy in vehicles and portable lights (flashlights and lanterns), starting emergency diesel generators, power sources in material handling equipment (forklifts), various portable electric/electronic equipment, backup emergency power for light-packs, hazard warning signal lights and flashlights, and standby power supplies or uninterruptible power supplies (UPS) for computer systems. In some instances, they are used as the only source of power; while in others (as mentioned above) they are used as a secondary or standby power supply. Currently, because of the oil crisis (high gasoline prices, etc.), hybrid battery-powered vehicles are receiving increased attention.

THE VOLTAIC CELL

The simplest cell (a device that transforms chemical energy into electrical energy) is known as a *voltaic* (or galvanic) cell (see figure 9.13). It consists of a piece of carbon (C) and a piece of zinc (Zn) suspended in a jar that contains a solution of water (H_2O) and sulfuric acid (H_2SO_4).

✔ **Note:** A simple cell consists of two strips, or *electrodes*, placed in a container that holds the *electrolyte*. A battery is formed when two or more cells are connected.

The electrodes are the conductors by which the current leaves or returns to the electrolyte. In the simple cell described above, they are carbon and zinc strips placed in the electrolyte. Zinc contains an abundance of negatively charged atoms, while

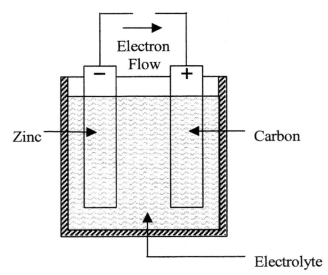

Figure 9.13.   Simple voltaic cell.

carbon has an abundance of positively charged atoms. When the plates of these materials are immersed in an electrolyte, chemical action between the two begins.

In the *dry cell* (see figure 9.14), the electrodes are the carbon rod in the center and the zinc container in which the cell is assembled. The electrolyte is the solution that acts upon the electrodes that are placed in it. The electrolyte may be a salt, an acid, or an alkaline solution. In the simple voltaic cell and in the automobile storage

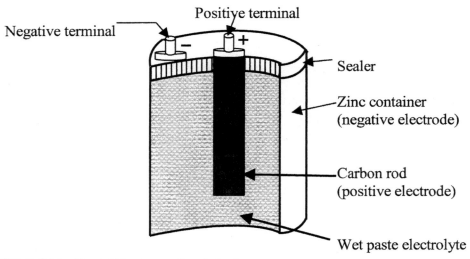

Figure 9.14.   Dry cell (cross-sectional view).

battery, the electrolyte is in a liquid form; in the dry cell, the electrolyte is a moist paste.

PRIMARY AND SECONDARY CELLS

Primary cells are normally those that cannot be recharged or returned to good condition after their voltage drops too low. Dry cells in flashlights and transistor radios are examples of primary cells. Some primary cells have been developed to the state where they can be recharged.

A *secondary cell* is one in which the electrodes and the electrolyte are altered by the chemical action that takes place when the cell delivers current. These cells are rechargeable. During recharging, the chemicals that provide electric energy are restored to their original condition. Recharging is accomplished by forcing an electric current through them in the opposite direction to that of discharge.

A cell is recharged by connecting as shown in figure 9.15. Some battery chargers have a voltmeter and an ammeter that indicate the charging voltage and current. The automobile storage battery is the most common example of the secondary cell.

BATTERY

As stated previously, a *battery* consists of two or more cells placed in a common container. The cells are connected in series, in parallel, or in some combination of series and parallel, depending upon the amount of voltage and current required of the battery.

Battery Operation

The chemical reaction within a battery provides the voltage. This occurs when a conductor is connected externally to the electrodes of a cell, causing electrons to flow under the influence of a difference in potential across the electrodes from the zinc

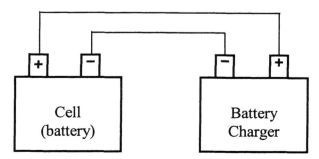

Figure 9.15.   Hookup for charging a secondary cell with a battery charger.

(negative) through the external conductor to the carbon (positive), returning within the solution to the zinc. After a short period of time, the zinc will begin to waste away because of the acid.

The voltage across the electrodes depends upon the materials from which the electrodes are made and the composition of the solution. The difference of potential between the carbon and zinc electrodes in a dilute solution of sulfuric acid and water is about 1.5 volts.

The current that a primary cell may deliver depends upon the resistance of the entire circuit, including that of the cell itself. The internal resistance of the primary cell depends upon the size of the electrodes, the distance between them in the solution, and the resistance of the solution. The larger the electrodes and the closer together they are in solution (without touching), the lower the internal resistance of the primary cell and the more current it is capable of supplying to the load.

▶ **Note:** When current flows through a cell, the zinc gradually dissolves in the solution and the acid is neutralized.

Combining Cells

In many operations, battery-powered devices may require more electrical energy than one cell can provide. Various devices may require either a higher voltage or more current, and in some cases both. Under such conditions it is necessary to combine, or interconnect, a sufficient number of cells to meet the higher requirements. Cells connected in series provide a higher voltage, while cells connected in parallel provide a higher current capacity. To provide adequate power when both voltage and current requirements are greater than the capacity of one cell, a combination series-parallel network of cells must be interconnected.

When cells are connected in *series* (see figure 9.16), the total voltage across the battery of cells is equal to the sum of the voltage of each of the individual cells. In figure 9.16, the four 1.5 V cells in series provide a total battery voltage of 6 V. When cells are placed in series, the positive terminal of one cell is connected to the negative terminal of the other cell. The positive electrode of the first cell and the negative electrode of the last cell then serve as the power takeoff terminals of the battery. The current flowing through such a battery of series cells is the same as from one cell because the same current flows through all the series cells.

To obtain a greater current, a battery has cells connected in *parallel*, as shown in figure 9.17. In this parallel connection, all the positive electrodes are connected to one line, and all negative electrodes are connected to the other line. Any point on the positive side can serve as the positive terminal of the battery and any point on the negative side can be the negative terminal.

The total voltage output of a battery of three parallel cells is the same as that for a single cell (figure 9.17), but the available current is three times that of one cell—that is, the current capacity has been increased.

Identical cells in parallel all supply equal parts of the current to the load. For

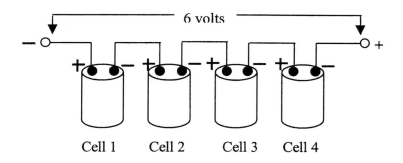

6 volts

− ○   ○ +

Cell 1   Cell 2   Cell 3   Cell 4

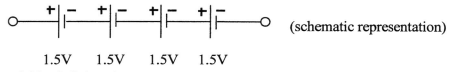

(schematic representation)

1.5V   1.5V   1.5V   1.5V

Figure 9.16.   Cells in series.

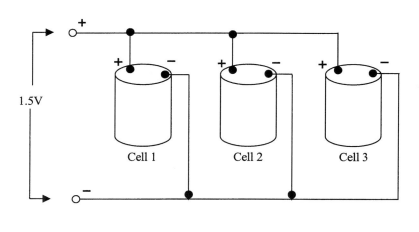

+

1.5V

Cell 1   Cell 2   Cell 3

−

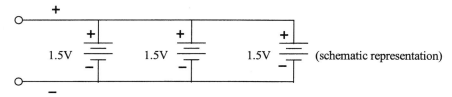

+

1.5V   1.5V   1.5V   (schematic representation)

−

Figure 9.17.   Cells in parallel.

example, of three different parallel cells producing a load current of 210 ma, each cell contributes 70 ma.

Figure 9.18 depicts a schematic of a *series-parallel* battery network supplying power to a load requiring both a voltage and current greater than one cell can provide. To provide the required increased voltage, groups of three 1.5-V cells are connected in series. To provide the required increased amperage, four series groups are connected in parallel.

Battery Characteristics

Batteries are generally classified by their various characteristics. Parameters such as internal resistance, specific gravity, capacity, and shelf life are used to classify batteries by type.

Regarding *internal resistance*, it is important to keep in mind that a battery is a d.c. voltage generator. As such, the battery has internal resistance. In a chemical cell, the resistance of the electrolyte between the electrodes is responsible for most of the cell's internal resistance. Because any current in the battery must flow through the internal resistance, this resistance is in series with the generated voltage. With no current, the voltage drop across the resistance is zero, so that the full generated voltage develops across the output terminals. This is the open-circuit voltage, or no-load voltage. If a load resistance is connected across the battery, the load resistance is in series with internal resistance. When current flows in this circuit, the internal voltage drop decreases the terminal voltage of the battery.

The ratio of the weight of a certain volume of liquid to the weight of the same volume of water is called the *specific gravity* of the liquid. Pure sulfuric acid has a specific gravity of 1.835, since it weighs 1.835 times as much as water per unit volume. The specific gravity of a mixture of sulfuric acid and water varies with the strength of the solution from 1.000 to 1.830.

The specific gravity of the electrolyte solution in a lead-acid cell ranges from 1.210 to 1.300 for new, fully charged batteries. The higher the specific gravity, the less internal resistance of the cell and the higher the possible load current. As the cell discharges, the water formed dilutes the acid and the specific gravity gradually decreases to about 1.150, at which time the cell is considered to be fully discharged.

The specific gravity of the electrolyte is measured with a *hydrometer*, which has a

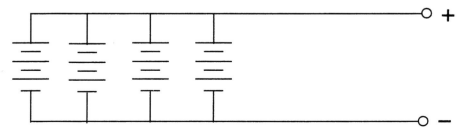

Figure 9.18.   Series-parallel connected cells.

compressible rubber bulb at the top, a glass barrel, and a rubber hose at the bottom of the barrel. In taking readings with a hydrometer, the decimal point is usually omitted. For example, a specific gravity of 1.260 is read simply as "twelve-sixty." A hydrometer reading of 1.210 to 1.300 indicates full charge; about 1.250 is half-charge; and 1.150 to 1.200 is complete discharge.

The *capacity* of a battery is measured in ampere-hours (Ah).

✔ **Note:** The ampere-hour capacity is equal to the product of the current in amperes and the time in hours during which the battery is supplying this current. The ampere-hour capacity varies inversely with the discharge current. The size of a cell is determined generally by its ampere-hour capacity.

The capacity of a storage battery determines how long it will operate at a given discharge rate and depends upon many factors. The most important of these are as follows:

- the area of the plates in contact with the electrolyte
- the quantity and specific gravity of the electrolyte
- the type of separators
- the general condition of the battery (degree of sulfating, plates bucked, separators warped, sediment in bottom of cells, etc.)
- the final limiting voltage

The *shelf life* of a cell is the period of time during which the cell can be stored without losing more than approximately 10% of its original capacity. The loss of capacity of a stored cell is due primarily to the drying out of its electrolyte in a wet cell and to chemical actions that change the materials within the cell. The shelf life of a cell can be extended by keeping it in a cool, dry place.

Conductors

Electric current moves easily through some materials but with greater difficulty through others. Three good electrical conductors are copper, silver, and aluminum (generally, we can say that most metals are good conductors). At the present time, copper is the material of choice used in electrical conductors. Under special conditions, certain gases are also used as conductors (e.g., neon gas, mercury vapor, and sodium vapor are used in various kinds of lamps).

The function of the wire conductor is to connect a source of applied voltage to a load resistance with a minimum IR voltage drop in the conductor, so that most of the applied voltage can produce current in the load resistance. Ideally, a conductor must have a very low resistance (e.g., a typical value for a conductor—copper—is less than 1Ω per 10 ft).

CONDUCTORS, SEMICONDUCTORS, AND INSULATORS

Substances that permit the free movement of a large number of electrons are called *conductors*. The most widely used electrical conductor is copper because of its high conductivity (how good a conductor the material is) and cost-effectiveness.

Electrical energy is transferred through a copper or other metal conductor by means of the movement of free electrons that migrate from atom to atom inside the conductor (see figure 9.3). Each electron moves a very short distance to the neighboring atom, where it replaces one or more electrons by forcing them out of their orbits. The replaced electrons repeat the process in other nearby atoms until the movement is transmitted throughout the entire length of the conductor. A good conductor is said to have a low opposition, or *resistance*, to the electron (current) flow.

✔ Note: If lots of electrons flow through a material with only a small force (voltage) applied, we call that material a *conductor*.

Table 9.1 lists many of the metals commonly used as electric conductors. The best conductors appear at the top of the list, with the poorer ones shown last.

✔ Note: The movement of each electron (e.g., in copper wire) takes a very small amount of time, almost instantly. This is an important point to keep in mind later in the book, when events in an electrical circuit seem to occur simultaneously.

While it is true that electron motion is known to exist to some extent in all matter, some substances, such as rubber, glass, and dry wood, have very few free electrons. In these materials large amounts of energy must be expended in order to break the electrons loose from the influence of the nucleus. Substances containing very few free electrons are called *insulators*. Insulators are important in electrical work because they prevent the current from being diverted from the wires.

✔ Note: If the voltage is large enough, even the best insulators will break down and allow their electrons to flow.

Table 9.2 lists some materials that we often use as insulators in electrical circuits. The list is in decreasing order of ability to withstand high voltages without conducting.

Table 9.1   Electrical conductors

Silver	Brass
Copper	Iron
Gold	Tin
Aluminum	Mercury
Zinc	

Table 9.2   Common insulators

Rubber	Plastics
Mica	Glass
Wax or paraffin	Fiberglass
Porcelain	Dry wood
Bakelite	Air

A material that is neither a good conductor nor a good insulator is called a *semi-conductor*. Silicon and germanium are substances that fall into this category. Because of their peculiar crystalline structure, these materials may under certain conditions act as conductors; under other conditions, as insulators. As the temperature is raised, however, a limited number of electrons become available for conduction.

UNIT SIZE OF CONDUCTORS

A standard (or unit size) of a conductor has been established to compare the resistance and size of one conductor with another. The unit of linear measurement used (in regard to diameter of a piece of wire) is the *mil* (0.001 of an inch). A convenient unit of wire length used is the foot. Thus, the standard unit of size in most cases is the *mil-foot* (i.e., a wire will have unit size if it has a diameter of 1 mil and a length of 1 foot). The resistance in ohms of a unit conductor or a given substance is called the *resistivity* (or specific resistance) of the substance.

As a further convenience, *gage* numbers are also used in comparing the diameter of wires. The B and S (Browne and Sharpe) gage was used in the past; now the most commonly used gage is the American Wire Gage (AWG).

Square Mil

Figure 9.19 (A) shows a square mil. The *square mil* is a convenient unit of cross-sectional area for square or rectangular conductors. As shown in figure 9.19 (A), a square mil is the area of a square, the sides of which are 1 mil. To obtain the cross-

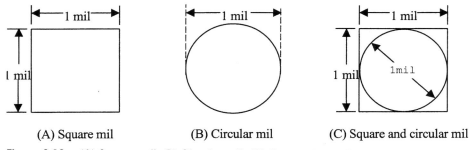

(A) Square mil   (B) Circular mil   (C) Square and circular mil

Figure 9.19.   (A) Square mil. (B) Circular mil. (C) Comparison of circular to square mil.

sectional area in square mils of a square conductor, square one side measured in mils. To obtain the cross-sectional area in square mils of a rectangular conductor, multiply the length of one side by that of the other, each length being expressed in mils.

Example 9.1
Problem
Find the cross-sectional area of a large rectangular conductor 5/8-inch thick and 5 inches wide.

Solution
The thickness may be expressed in mils as $0.625 \times 1,000 = 625$ mils and the width as $5 \times 1,000 = 5,000$ mils. The cross-sectional area is $625 \times 5,000$, or 3,125,000 square mils.

Circular Mil

The *circular mil* is the standard unit of wire cross-sectional area used in most wire tables. To avoid the use of decimals (because most wires used to conduct electricity may be only a small fraction of an inch), it is convenient to express these diameters in mils. For example, the diameter of a wire is expressed as 25 mils instead of 0.025 inch. A circular mil is the area of a circle having a diameter of 1 mil, as shown in figure 9.19 (B). The area in circular mils of a round conductor is obtained by squaring the diameter measured in mils. Thus, a wire having a diameter of 25 mils has an area of 25^2 or 625 circular mils. By way of comparison, the basic formula for the area of a circle is

$$A = \pi R^2 \qquad (9.1)$$

and in this example the area in square inches is

$$A = \pi R^2 = 3.14\,(0.0125)^2 = 0.00049 \text{ sq in.}$$

If D is the diameter of a wire in mils, the area in square mils can be determined using

$$A = \pi\,(D/2)^2 \qquad (9.2)$$

which translates to

$$= 3.14/4\ D^2$$
$$= 0.785\ D^2 \text{ sq mils}$$

Thus, a wire 1 mil in diameter has an area of

$$A = 0.785 \times 1^2 = 0.785 \text{ sq mils}$$

which is equivalent to 1 circular mil. The cross-sectional area of a wire in circular mils is therefore determined as

$$A = \frac{0.785 \, D^2}{0.785} = D^2 \text{ circular mils,}$$

where D is the diameter in mils. Therefore, the constant $\pi/4$ is eliminated from the calculation.

It should be noted that in comparing square and round conductors that the circular mil is a smaller unit of area than the square mil, and therefore there are more circular mils than square mils in any given area. The comparison is shown in figure 9.19 (C). The area of a circular mil is equal to 0.785 of a square mil.

✔ **Important Point**: To determine the circular-mil area when the square-mil area is given, divide the area in square mils by 0.785. Conversely, to determine the square-mil area when the circular-mil area is given, multiply the area in circular mils by 0.785.

Example 9.2
Problem
A No. 12 wire has a diameter of 80.81 mils. What is (1) its area in circular mils and (2) its area in square mils?

Solution

$$\text{(1)} \quad A = D^2 = 80.81^2 = 6{,}530 \text{ circular mils}$$
$$\text{(2)} \quad A = 0.785 \times 6{,}530 = 5{,}126 \text{ square mils}$$

Example 9.3
Problem
A rectangular conductor is 1.5 inches wide and 0.25 inch thick. (1) What is its area in square mils? (2) What size of round conductor in circular mils is necessary to carry the same current as the rectangular bar?

Solution

(1) 1.5″ = 1.5 × 1,000 = 1,500 mils
   0.25″ = 0.25 × 1,000 = 250 mils
   A = 1,500 × 250 = 375,000 sq mils
(2) To carry the same current, the cross-sectional area of the rectangular bar and the cross-sectional area of the round conductor must be equal. There are more circular mils than square mils in this area, and therefore

$$A = \frac{375{,}000}{0.785} = 477{,}700 \text{ circular mils}$$

Circular-Mil-Foot

As shown in figure 9.20, a *circular-mil foot* is actually a unit of volume. More specifically, it is a unit conductor 1 foot in length and having a cross-sectional area of 1

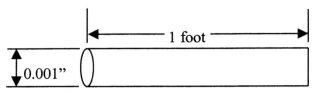

Figure 9.20.   Circular-mil foot.

circular mil. Because it is considered a unit conductor, the circular-mil-foot is useful in making comparisons between wires that are made of different metals. For example, a basis of comparison of the resistivity of various substances may be made by determining the resistance of a circular-mil-foot of each of the substances.

Resistivity

All materials differ in their atomic structure and therefore in their ability to resist the flow of an electric current. The measure of the ability of a specific material to resist the flow of electricity is called its *resistivity*, or specific resistance—the resistance in ohms offered by unit volume (the circular-mil-foot) of a substance to the flow of electric current. Resistivity is the reciprocal of conductivity (i.e., the ease by which current flows in a conductor). A substance that has a high resistivity will have a low conductivity, and vice versa.

The resistance of a given length, for any conductor, depends upon the resistivity of the material, the length of the wire, and the cross-sectional area of the wire according to the equation

$$R = \rho \frac{L}{A} \tag{9.3}$$

where

R = resistance of the conductor, Ω
L = length of the wire, ft
A = cross-sectional area of the wire, CM
ρ = specific resistance or resistivity, CM $\times$ Ω/ft

The factor ρ (Greek letter rho, pronounced "roe") permits different materials to be compared for resistance according to their nature without regard to different lengths or areas. Higher values of ρ mean more resistance.

✔ **Key Point:** The resistivity of a substance is the resistance of a unit volume of that substance.

Many tables of resistivity are based on the resistance in ohms of a volume of the substance 1 foot long and 1 circular mil in cross-sectional area. The temperature at which the resistance measurement is made is also specified. If the kind of metal of which the conductor is made is known, the resistivity of the metal may be obtained from a table. The resistivity, or specific resistance, of some common substances is given in table 9.3.

Example 9.4
Problem
What is the resistance of 1,000 feet of copper wire having a cross-sectional area of 10,400 circular mils (No. 10 wire), the wire temperature being 20°C?

Solution
The resistivity (specific resistance), from table 9.3, is 10.37. Substituting the known values in the preceding equation (4.3), the resistance, R, is determined as

$$R = \rho \frac{L}{A} = 10.37 \times \frac{1,000}{10,400} = 1 \text{ ohm, approximately}$$

Temperature Coefficient

The resistance of pure metals, such as silver, copper, and aluminum, increases as the temperature increases. The *temperature coefficient* of resistance, α (Greek letter alpha), indicates how much the resistance changes for a change in temperature. A positive value for α means R increases with temperature; a negative α means R decreases; and a zero α means R is constant, not varying with changes in temperature. Typical values of α are listed in table 9.4.

    The amount of increase in the resistance of a 1-ohm sample of the copper conductor per degree rise in temperature (i.e., the temperature coefficient of resistance) is approximately 0.004. For pure metals, the temperature coefficient of resistance ranges between 0.004 and 0.006 ohm.

    Thus, a copper wire having a resistance of 50 ohms at an initial temperature of 0°C will have an increase in resistance of 50 × 0.004, or 0.2 ohms (approximate) for the entire length of wire for each degree of temperature rise above 0°C. At 20°C the increase in resistance is approximately 20 × 0.2, or 4 ohms. The total resistance at 20°C is 50 + 4, or 54 ohms.

Table 9.3   Resistivity (specific resistance)

Substance	Specific resistance @ 20° (CM ft (Ω))
Silver	9.8
Copper (drawn)	10.37
Gold	14.7
Aluminum	17.02
Tungsten	33.2
Brass	42.1
Steel (soft)	95.8
Nichrome	660.0

Table 9.4   Properties of conducting materials (approximate)

Material	Temperature coefficient, $\Omega/^{\circ}C$
Aluminum	0.004
Carbon	−0.0003
Constantan	0 (average)
Copper	0.004
Gold	0.004
Iron	0.006
Nichrome	0.0002
Nickel	0.005
Silver	0.004
Tungsten	0.005

✔ **Note:** As shown in table 9.4, carbon has a negative temperature coefficient. In general, α is negative for all semiconductors such as germanium and silicon. A negative value for α means less resistance at higher temperatures. Therefore, the resistance of semiconductor diodes and transistors can be reduced considerably when they become hot with normal load current. Observe, also, that constantan has a value of zero for α (table 9.4). Thus, it can be used for precision wire-wound resistors, which do not change resistance when the temperature increases.

Conductor Insulation

Electric current must be contained; it must be channeled from the power source to a useful load—safely. To accomplish this, electric current must be forced to flow only where it is needed. Moreover, current-carrying conductors must not be allowed (generally) to come in contact with one another, their supporting hardware, or personnel working near them. To accomplish this, conductors are coated or wrapped with various materials. These materials have such a high resistance that they are, for all practical purposes, nonconductors. They are generally referred to as *insulators* or *insulating materials*.

There are a wide variety of insulated conductors available to meet the requirements of any job. However, only the necessary minimum of insulation is applied for any particular type of cable designed to do a specific job. This is the case because insulation is expensive and has a stiffening effect, and is required to meet a great variety of physical and electrical conditions.

Two fundamental but distinctly different properties of insulation materials (e.g., rubber, glass, asbestos, and plastics) are insulation resistance and dielectric strength.

1. *Insulation resistance* is the resistance to current leakage through and over the surface of insulation materials.
2. *Dielectric strength* is the ability of the insulator to withstand potential difference and is usually expressed in terms of the voltage at which the insulation fails because of the electrostatic stress.

Various types of materials are used to provide insulation for electric conductors, including rubber, plastic, varnished cloth, paper, silk and cotton, and enamel.

Electromagnetism

In medicine, anatomy and physiology are so closely related that the medical student cannot study one at length without involving the other. A similar relationship holds for the electrical field; that is, magnetism and basic electricity are so closely related that one cannot be studied at length without involving the other. This close fundamental relationship is continually borne out in subsequent sections of this chapter, such as in the study of generators, transformers, and motors. To be proficient in the knowledge of electricity, one must become familiar with such general relationships that exist between magnetism and electricity as the following:

- Electric current flow will always produce some form of magnetism.
- Magnetism is by far the most commonly used means for producing or using electricity.
- The peculiar behavior of electricity under certain conditions is caused by magnetic influences.

MAGNETIC FIELD AROUND A SINGLE CONDUCTOR

In 1819, Hans Christian Oersted, a Danish scientist, discovered that a field of magnetic force exists around a single wire conductor carrying an electric current. In figure 9.21, a wire is passed through a piece of cardboard and connected through a switch to a dry cell. With the switch open (no current flowing), if we sprinkle iron filings on the cardboard and tap it gently, the filings will fall back haphazardly. Now, if we close the switch, current will begin to flow in the wire. If we tap the cardboard again, the magnetic effect of the current in the wire will cause the filings to fall back into a definite pattern of concentric circles with the wire as the center of the circles. Every section of the wire has this field of force around it in a plane perpendicular to the wire, as shown in figure 9.22.

The ability of the magnetic field to attract bits of iron (as demonstrated in figure 9.21) depends on the number of lines of force present. The strength of the magnetic field around a wire carrying a current depends on the current, since it is the current that produces the field. The greater the current, the greater the strength of the field. A large current will produce many lines of force extending far from the wire, while a small current will produce only a few lines close to the wire, as shown in figure 9.23.

POLARITY OF A SINGLE CONDUCTOR

The relation between the direction of the magnetic lines of force around a conductor and the direction of current flow along the conductor may be determined by means

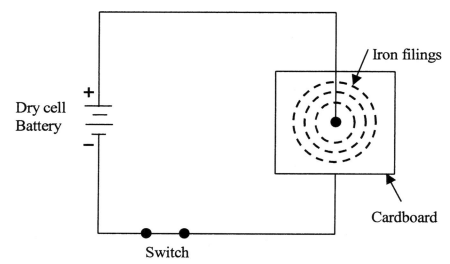

Figure 9.21.   A circular patterns of magnetic force exists around a wire carrying an electric current.

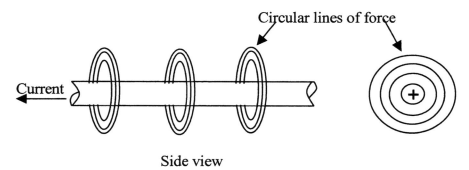

Side view

Figure 9.22.   The circular fields of force around a wire carrying a current are in planes that are perpendicular to the wire.

Figure 9.23.   The strength of the magnetic field around a wire carrying a current depends on the amount of current.

of the *left-hand rule for a conductor.* If the conductor is grasped in the left hand with the thumb extended in the direction of electron flow (− to +), the fingers will point in the direction of the magnetic lines of force. This is the same direction that the north pole of a compass would point if the compass were placed in the magnetic field.

✔ **Important Note:** Arrows are generally used in electric diagrams to denote the direction of current flow along the length of wire. Where cross sections of wire are shown, a special view of the arrow is used. A cross-sectional view of a conductor that is carrying current toward the observer is illustrated in figure 9.24 (A). The direction of current is indicated by a dot, which represents the head of the arrow. A conductor that is carrying current away from the observer is illustrated in figure 9.24 (B). The direction of current is indicated by a cross, which represents the tail of the arrow.

FIELD AROUND TWO PARALLEL CONDUCTORS

When two parallel conductors carry current in the same direction, the magnetic fields tend to encircle both conductors, drawing them together with a force of attraction, as shown in figure 9.25 (A). Two parallel conductors carrying currents in opposite directions are shown in figure 9.25 (B). The field around one conductor is opposite in direction to the field around the other conductor. The resulting lines of force are crowded together in the space between the wires, and tend to push the wires apart. Therefore, two parallel adjacent conductors carrying currents in the same direction attract each other, and two parallel conductors carrying currents in opposite directions repel each other.

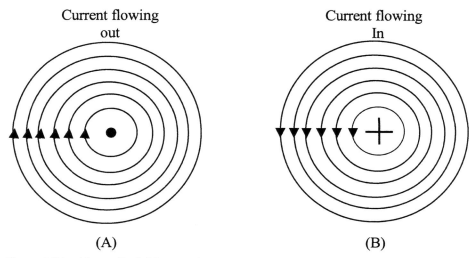

Figure 9.24. Magnetic field around a current-carrying conductor.

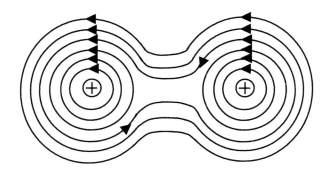

(A) Current flowing in the same direction

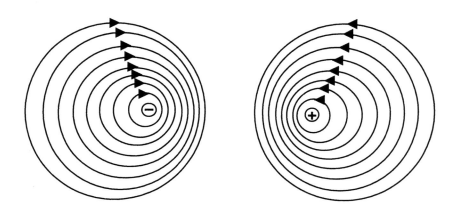

(B) Currents flowing in opposite directions

Figure 9.25.   Magnetic field around two parallel conductors.

MAGNETIC FIELD OF A COIL

The magnetic field around a current-carrying wire exists at all points along its length. Bending the current-carrying wire into the form of a single loop has two results. First, the magnetic field consists of more dense concentric circles in a plane perpendicular to the wire (see figure 9.22), although the total number of lines is the same as for the straight conductor. Second, all the lines inside the loop are in the same direction. When this straight wire is wound around a core, as is shown in figure 9.26, it becomes a coil and the magnetic field assumes a different shape. When current is passed through the coiled conductor, the magnetic field of each turn of wire links with the fields of adjacent turns. The combined influence of all the turns produces a two-pole field similar to that of a simple bar magnet. One end of the coil will be a north pole and the other end will be a south pole.

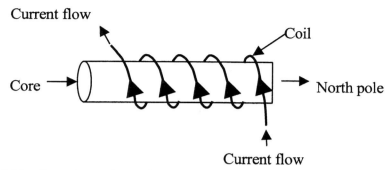

Figure 9.26.   Current-carrying coil.

POLARITY OF AN ELECTROMAGNETIC COIL

In figure 9.23, it was shown that the direction of the magnetic field around a straight conductor depends on the direction of current flow through that conductor. Thus, a reversal of current flow through a conductor causes a reversal in the direction of the magnetic field that is produced. It follows that a reversal of the current flow through a coil also causes a reversal of its two-pole field. This is true because that field is the product of the linkage between the individual turns of wire on the coil. Therefore, if the field of each turn is reversed, it follows that the total field (coil's field) is also reversed.

When the direction of electron flow through a coil is known, its polarity may be determined by use of the *left-hand rule for coils*. This rule is illustrated in figure 9.26, and is stated as follows: Grasping the coil in the left hand, with the fingers "wrapped around" in the direction of electron flow, the thumb will point toward the north pole.

STRENGTH OF AN ELECTROMAGNETIC FIELD

The strength, or intensity, of the magnetic field of a coil depends on a number of factors.

• the *number of turns* of conductor
• the *amount of current flow* through the coil
• the *ratio of the coil's length to its width*
• the *type of material in the core*

MAGNETIC UNITS

The law of current flow in the electric circuit is similar to the law for the establishing of flux in the magnetic circuit.

The *magnetic flux*, φ (phi), is similar to current in the Ohm's law formula, and comprises the total number of lines of force existing in the magnetic circuit. The *maxwell* is the unit of flux—that is, 1 line of force is equal to 1 maxwell.

✔ **Note:** The maxwell is often referred to as simply a line of force, line of induction, or line.

The *strength* of a magnetic field in a coil of wire depends on how much current flows in the turns of the coil. The more current, the stronger the magnetic field. Also, the more turns, the more concentrated are the lines of force. The force that produces the flux in the magnetic circuit (comparable to electromotive force in Ohm's law) is known as *magnetomotive force*, or mmf. The practical unit of magnetomotive force is the *ampere-turn* (At). In equation form,

$$F = \text{ampere-turns} = NI \tag{9.4}$$

where

F = magnetomotive force, At
N = number of turns
I = current, A

Example 9.5
Problem
Calculate the ampere-turns for a coil with 2000 turns and a 5-Ma current.

Solution
Use equation (9.4) and substitute N = 2000 and I = 5×10^{-3} A.

$$NI = 2000 \, (5 \times 10^{-3}) = 10 \text{ At}$$

The unit of *intensity* of magnetizing force per unit of length is designated as H, and is sometimes expressed as gilberts per centimeter of length. Expressed as an equation,

$$H = \frac{NI}{L} \tag{9.5}$$

where

H = magnetic field intensity, ampere-turns per meter (At/m)
NI = ampere-turns, At
L = length between poles of the coil, m

✔ **Note:** Equation 9.5 is for a solenoid. H is the intensity of an air core. With an iron core, H is the intensity through the entire core, and L is the length or distance between poles of the iron core.

PROPERTIES OF MAGNETIC MATERIALS

In this section we discuss two important properties of magnetic materials: permeability and hysteresis.

Permeability

When the core of an electromagnet is made of annealed sheet steel it produces a stronger magnet than if a cast iron core is used. This is the case because annealed sheet steel is more readily acted upon by the magnetizing force of the coil than is the hard cast iron. Simply put, soft sheet steel is said to have greater *permeability* because of the greater ease with which magnetic lines are established in it.

Recall that permeability is the relative ease with which a substance conducts magnetic lines of force. The permeability of air is arbitrarily set at 1. The permeability of other substances is the ratio of their ability to conduct magnetic lines compared to that of air. The permeability of nonmagnetic materials, such as aluminum, copper, wood, and brass is essentially unity, or the same as for air.

✔ **Important Note:** The permeability of magnetic materials varies with the degree of magnetization, being smaller for high values of flux density.

✔ **Important Note:** *Reluctance*, which is analogous to resistance, is the opposition to the production of flux in a material; it is inversely proportional to permeability. Iron has high permeability and, therefore, low reluctance. Air has low permeability and hence high reluctance.

Hysteresis

When the current in a coil of wire reverses thousands of times per second, a considerable loss of energy can occur. This loss of energy is caused by *hysteresis*. Hysteresis means "a lagging behind"—that is, the magnetic flux in an iron core lags behind the increases or decreases of the magnetizing force.

The simplest method of illustrating the property of hysteresis is by graphical means, such as the hysteresis loop shown in figure 9.27.

The hysteresis loop is a series of curves that show the characteristics of a magnetic material. Opposite directions of current result are in the opposite directions of $+H$ and $-H$ for field intensity. Similarly, opposite polarities are shown for flux density as $+B$ and $-B$. The current starts at the center (zero) when the material is unmagnetized. Positive H values increase B to saturation at $+B_{max}$. Next H decreases to zero, but B drops to the value of B, because of hysteresis. The current that produced the original magnetization now is reversed so that H becomes negative. B drops to zero and continues to $-B_{max}$. As the $-H$ values decrease, B is reduced to $-B$, when H is zero. Now with a positive swing of current, H becomes positive, producing saturation at $+B_{max}$ again. The hysteresis loop is now completed. The curve doesn't return to zero at the center because of hysteresis.

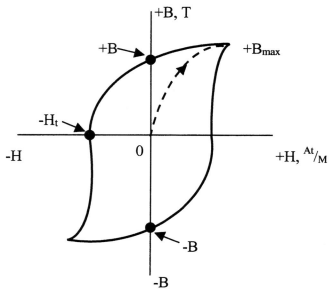

Figure 9.27.   Hysteresis loop.

ELECTROMAGNETS

An *electromagnet* is composed of a coil of wire wound around a core that is normally soft iron, because of its high permeability and low hysteresis. When direct current flows through the coil, the core will become magnetized with the same polarity that the coil would have without the core. If the current is reversed, the polarity of the coil and core are reversed.

The electromagnet is of great importance in electricity simply because the magnetism can be "turned on" or "turned off" at will. The starter solenoid (an electromagnet) in automobiles and power boats is a good example. In an automobile or boat, an electromagnet is part of a relay that connects the battery to the induction coil, which generates the very high voltage needed to start the engine. The starter solenoid isolates this high voltage from the ignition switch. When no current flows in the coil, it is an "air core," but when the coil is energized, a movable soft-iron core does two things. First, the magnetic flux is increased because the soft-iron core is more permeable than the air core. Second, the flux is more highly concentrated. All this concentration of magnetic lines of force in the soft-iron core results in a very good magnet when current flows in the coil. But soft-iron loses its magnetism quickly when the current is shut off. The effect of the soft iron is, of course, the same whether it is movable, as in some solenoids, or permanently installed in the coil. An electromagnet then consists basically of a coil and a core; it becomes a magnet when current flows through the coil.

The ability to control the action of magnetic force makes an electromagnet very useful in many circuit applications. Many of the applications of electromagnets are discussed throughout this chapter.

Simple Electric Circuit

A simple electric circuit includes: an energy source (difference of potential; source of electromotive force [emf]; voltage), a conductor (wire), a load, and a means of control (see figure 9.28). The energy source could be a battery, as in figure 9.28, or some other means of producing a voltage. The load that dissipates the energy could be a lamp, a resistor, or a device that does useful work, such as an electric toaster, a computer, a power drill, a radio, or a soldering iron. Conductors are wires that offer low resistance to current; they connect all the loads in the circuit to the voltage source. No electrical device dissipates energy unless current (electrons) flows through it. Because conductors or wires are not perfect conductors, they heat up (dissipate energy), so they are actually part of the load. For simplicity, however, we usually think of the connecting wiring as having no resistance, since it would be tedious to assign a very low resistance value to the wires every time we wanted to calculate the solution to a problem. Control devices might be switches, variable resistors, circuit breakers, fuses, or relays.

A complete pathway for current flow, or closed circuit (figure 9.28), is an unbroken path for current from the emf, through a load, and back to the source. A circuit is called *open* (see figure 9.29) if a break in the circuit (e.g., an open switch) does not provide a complete path for current flow.

✔ **Important Point**: Current flows from the negative ($-$) terminal of the battery, shown in figures 9.28 and 9.29, through the load to the positive ($+$) battery terminal, and continues by going through the battery from the positive ($+$) terminal to the negative ($-$) terminal.

To protect a circuit, a fuse or circuit breaker is placed directly into the circuit (see figure 9.30). A fuse will open the circuit whenever a dangerously large current starts

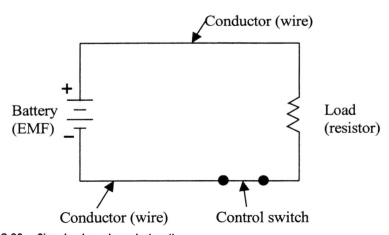

Figure 9.28.   Simple d.c. closed circuit.

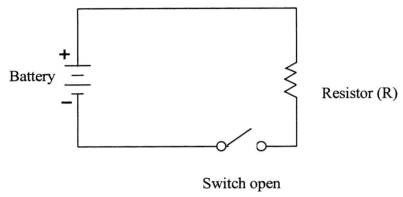

Switch open

Figure 9.29.   Open circuit.

to flow (i.e., a short-circuit condition occurs, caused by an accidental connection between two points in a circuit that offers very little resistance). A fuse will permit currents smaller than the fuse value to flow but will melt and therefore break or open the circuit if a larger current flows.

SCHEMATIC REPRESENTATION

The simple circuits shown in figures 9.28, 9.29, and 9.30 are displayed in schematic form. A *schematic diagram* (usually shortened to *schematic*) is a simplified drawing that represents the electrical, not the physical, situation in a circuit. The symbols used in schematic diagrams are the electrician's shorthand; they make the diagrams easier to draw and easier to understand. Consider the symbol in figure 9.31 used to represent a battery power supply. The symbol is rather simple and straightforward—but is also very important. For example, by convention, the shorter line in the symbol for a

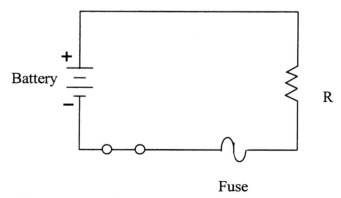

Fuse

Figure 9.30.   Simple fused circuit.

battery represents the negative terminal. It is important to remember this, because it is sometimes necessary to note the direction of current flow, which is from negative to positive, when you examine the schematic. The battery symbol shown in figure 9.31 has a single cell, so only one short and one long line are used. The number of lines used to represent a battery vary (and they are not necessarily equivalent to the number of cells), but they are always in pairs, with long and short lines alternating. In the circuit shown in figure 9.30, the current would flow in a counterclockwise direction—that is, in the opposite direction that a clock's hands move. If the long and short lines of the battery symbol (shown in figure 9.31) were reversed, the current in the circuit shown in figure 9.30 would flow clockwise—that is, in the direction of a clock's hands.

✔ **Important Note:** In studies of electricity and electronics, many circuits are analyzed that consist mainly of specially designed resistive components. As previously stated, these components are called resistors. Throughout the remaining analysis of the basic circuit, the resistive component will be a physical resistor. However, the resistive component could be any one of several electrical devices.

Keep in mind that in the simple circuits shown in the figures to this point we have only illustrated and discussed a few of the many symbols used in schematics to represent circuit components. (Other symbols will be introduced as we need them.)

It is also important to keep in mind that a closed loop of wire (conductor) is not necessarily a circuit. A source of voltage must be included to make it an electric circuit. In any electric circuit where electrons move around a closed loop, current, voltage, and resistance are present. The physical pathway for current flow is actually the circuit. By knowing any two of the three quantities, such as voltage and current, the third (resistance) may be determined. This is done mathematically using Ohm's law, which is the foundation on which electrical theory is based.

OHM'S LAW

Simply put, *Ohm's law* defines the relationship between current, voltage, and resistance in electric circuits. Ohm's law can be expressed mathematically in three ways:

a. The *current* (I) in a circuit is equal to the voltage applied to the circuit divided by the resistance of the circuit. Stated another way, the current in a circuit is *directly*

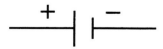

Figure 9.31. Schematic symbol for a battery.

proportional to the applied voltage and *inversely* proportional to the circuit resistance. Ohm's law may be expressed as an equation:

$$I = \frac{E}{R}$$

where

I = current in amperes (amps)
E = voltage in volts
R = resistance in ohms

b. The *resistance* (R) of a circuit is equal to the voltage applied to the circuit divided by the current in the circuit:

$$R = \frac{E}{I}$$

c. The applied *voltage* (E) to a circuit is equal to the product of the current and the resistance of the circuit:

$$E = I \times R = IR$$

If any two of the quantities in an equation are known, the third may be easily found. Let's look at an example.

Example 9.6
Problem
Figure 9.32 shows a circuit containing a resistance of 6 ohms and a source of voltage of 3 volts. How much current flows in the circuit?

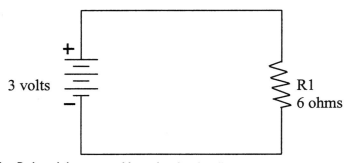

Figure 9.32. Determining current in a simple circuit.

Given:

E = 3 volts
R = 6 ohms
I = ?

Solution

$$I = \frac{E}{R}$$

$$I = \frac{3}{6}$$

$$I = 0.5 \text{ amperes}$$

To observe the effect of source voltage on circuit current, we use the circuit shown in figure 9.32 but double the voltage to 6 volts.

Example 9.7
Problem
Given:

E = 6 volts
R = 6 ohms
I = ?

Solution

$$I = \frac{E}{R}$$

$$I = \frac{6}{6}$$

$$I = 1 \text{ ampere}$$

Notice that as the source of voltage doubles, the circuit current also doubles.

✔ **Key Point:** Circuit current is directly proportional to applied voltage and will change by the same factor that the voltage changes.

To verify that current is inversely proportional to resistance, assume the resistor in figure 9.32 to have a value of 12 ohms.

Example 9.8
Problem
Given:

E = 3 volts
R = 12 ohms
I = ?

Solution

$$I = \frac{E}{R}$$

$$I = \frac{3}{12}$$

$$I = 0.25 \text{ amperes}$$

Comparing this current of 0.25 ampere for the 12-ohm resistor, to the 0.5-ampere of current obtained with the 6-ohm resistor, shows that doubling the resistance will reduce the current to one-half the original value. The point is that circuit current is inversely proportional to the circuit resistance.

Recall that if you know any two quantities E, I, and R, you can calculate the third. In many circuit applications current is known and either the voltage or the resistance will be the unknown quantity. To solve a problem in which current and resistance are known, the basic formula for Ohm's law must be transposed to solve for E, for I, or for R.

However, the Ohm's law equations can be memorized and practiced effectively by using an Ohm's law pie circle (see figure 9.33).

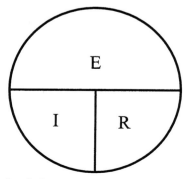

Figure 9.33. Ohm's Law pie circle.

Example 9.9
Problem
An electric light bulb draws 0.5 A when operating on a 120-V d.c. circuit. What is the resistance of the bulb?

Solution
The first step in solving a circuit problem is to sketch a schematic diagram of the circuit itself, labeling each of the parts and showing the known values (see figure 9.34).

Because I and E are known, we use

$$R = \frac{E}{I} = \frac{120}{0.5} = 240 \; \Omega$$

ELECTRIC POWER

Power, whether electrical or mechanical, pertains to the rate at which work is being done, so the power consumption in a facility is related to current flow. A large electric motor or air dryer consumes more power (and draws more current) in a given length of time than, for example, an indicating light on a motor controller. *Work* is done whenever a force causes motion. If a mechanical force is used to lift or move a weight, work is done. However, force exerted without causing motion, such as the force of a compressed spring acting between two fixed objects, does not constitute work.

✔ **Key Point:** Power is the rate at which work is done.

Electrical Power Calculations

The electric power P used in any part of a circuit is equal to the current I in that part multiplied by the V across that part of the circuit. In equation form,

$$P = I \times E \tag{9.6}$$

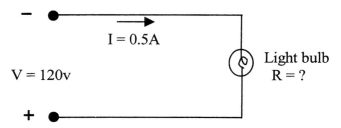

I = 0.5A

V = 120v

Light bulb
R = ?

Figure 9.34.   (For example 9.9.)

where

P = power, watts (W)
E = voltage, V
I = current, A

If we know the current I and the resistance R but not the voltage V, we can find the power P by using Ohm's law for voltage, so that substituting

$$E = IR$$

into equation 9.4, we have

$$P = IR \times I = I^2R \qquad (9.7)$$

In the same manner, if we know the voltage V and the resistance R but not the current I, we can find P by using Ohm's law for current, so that substituting

$$I = \frac{E}{R}$$

into equation 9.6, we have

$$P = E\frac{E}{R} = \frac{E^2}{R} \qquad (9.8)$$

✔ **Key Point:** If we know any two quantities, we can calculate the third.

Example 9.10
Problem
The current through a 200-Ω resistor to be used in a circuit is 0.25 A. Find the power rating of the resistor.

Solution
Since I and R are known, use

$$P = I^2R = (0.25)^2(200) = 0.0625(200) = 12.5 \text{ W}$$

✔ **Important Point:** The power rating of any resistor used in a circuit should be twice the wattage calculated by the power equation to prevent the resistor from burning out. Thus, the resistor used in example 9.10 should have a power rating of 25 watts.

Example 9.11
Problem
How many kilowatts of power are delivered to a circuit by a 220-V generator that supplies 30 A to the circuit?

Solution
Since V and I are given, use equation 3.4 to find P.

$$P = EI = 220(30) = 6600 \text{ W} = 6.6 \text{ Kw}$$

Example 9.12
Problem
If the voltage across a 30,000-Ω resistor is 450 V, what is the power dissipated in the resistor?

Solution
Since R and E are known, use

$$P = \frac{E^2}{R} = \frac{450^2}{30,000} = \frac{202,500}{30,000} = 6.75 \text{ W}$$

In this section, P was expressed in terms of alternate pairs of the other three basic quantities E, I, and R. In practice, you should be able to express any one of the three basic quantities, as well as P, in terms of any two of the others. Figure 9.35 is a summary of twelve basic Ohm's law formulas. The four quantities E, I, R, and P are at the center of the figure. Adjacent to each quantity are three segments. Note that in each segment, the basic quantity is expressed in terms of two other basic quantities, and no two segments are alike.

ELECTRIC ENERGY

Energy (the mechanical definition) is defined as the ability to do work (energy and time are essentially the same and are expressed in identical units). Energy is expended when work is done, because it takes energy to maintain a force when that force acts through a distance. The total energy expended to do a certain amount of work is equal to the working force multiplied by the distance through which the force moved to do the work.

In electricity, total energy expended is equal to the *rate* at which work is done, multiplied by the length of time the rate is measured. Essentially, energy W is equal to power P times time t.

The kilowatt-hour (Kwh) is a unit commonly used for large amounts of electric energy or work. The amount of kilowatt-hours is calculated as the product of the power in kilowatts (Kw) and the time in hours (h) during which the power is used.

$$\text{Kwh} = \text{Kw} \times \text{h} \qquad\qquad (9.9)$$

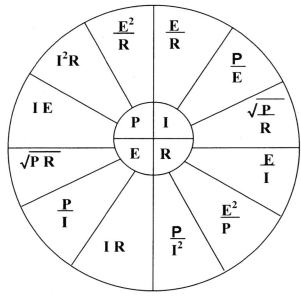

Figure 9.35.   Ohm's Law circle: summary of basic formulas.

Example 9.13
Problem
How much energy is delivered in 4 hours by a generator supplying 12 Kw?

Solution

$$\text{Kwh} = \text{Kw} \times \text{h}$$
$$= 12(4) = 48$$
$$\text{Energy delivered} = 48 \text{ kWh}$$

SERIES D.C. CIRCUITS

As previously mentioned, an electric circuit is made up of a voltage source, the necessary connecting conductors, and the effective load.

If the circuit is arranged so that the electrons have only one possible path, the circuit is called a *series circuit*. Therefore, a series circuit is defined as a circuit that contains only one path for current flow. Figure 9.36 shows a series circuit having several loads (resistors).

✔ **Key Point:** A *series circuit* is a circuit in which there is only one path for current to flow along.

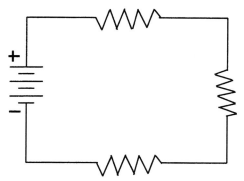

Figure 9.36. Series circuit.

Resistance in Series Circuit

Referring to figure 9.36, the current in a series circuit, in completing its electrical path, must flow through each resistor inserted into the circuit. Thus, each additional resistor offers added resistance. In a series circuit, *the total circuit resistance (R_T) is equal to the sum of the individual resistances.* As an equation:

$$R_T = R_1 + R_2 + R_3 \ldots R_n \qquad (9.10)$$

where

R_T = total resistance, Ω
R_1, R_2, R_3 = resistance in series, Ω
R_n = any number of additional resistors in equation

Example 9.14
Problem
Three resistors of 10 ohms, 12 ohms, and 25 ohms are connected in series across a battery whose emf is 110 volts (figure 9.37). What is the total resistance?

Solution
Given:

R_1 = 10 ohms
R_2 = 12 ohms
R_3 = 25 ohms
R_T = ?

$$R_T = R_1 + R_2 + R_3$$
$$R_T = 10 + 12 + 25$$
$$R_T = 47 \text{ ohms}$$

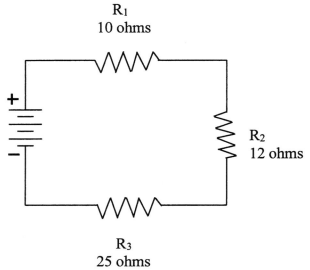

R₁
10 ohms

R₂
12 ohms

R₃
25 ohms

Figure 9.37.   Solving for total resistance in a series circuit.

Example 9.15
Problem
The total resistance of a circuit containing three resistors is 50 ohms (see figure 9.38).
Two of the circuit resistors are 12 ohms each. Calculate the value of the third resistor.

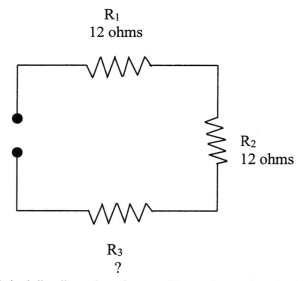

R₁
12 ohms

R₂
12 ohms

R₃
?

Figure 9.38.   Calculating the value of one resistance in a series circuit.

Solution
Given:

R_T = 50 ohms
R_1 = 12 ohms
R_2 = 12 ohms
R_3 = ?

$$R_T = R_1 + R_2 + R_3$$

Subtracting $(R_1 + R_2)$ from both sides of the equation

$$R_3 = R_T - R_1 - R_2$$
$$R_3 = 50 - 12 - 12$$
$$R_3 = 50 - 24$$
$$R_3 = 26 \text{ ohms}$$

✔ **Key Point:** When resistances are connected in series, the total resistance in the circuit is equal to the sum of the resistances of all the parts of the circuit.

Current in Series Circuit

Because there is but one path for current in a series circuit, the same current (I) must flow through each part of the circuit. Thus, to determine the current throughout a series circuit, only the current through one of the parts need be known.

The fact that the same current flows through each part of a series circuit can be verified by inserting ammeters into the circuit at various points as shown in figure 9.39. As indicated in figure 9.39, each meter indicates the same value of current.

✔ **Key Point:** In a series circuit the same current flows in every part of the circuit. *Do not* add the currents in each part of the circuit to obtain current (I).

Voltage in Series Circuit

The *voltage* drop across the resistor in the basic circuit is the total voltage across the circuit and is equal to the applied voltage. The total voltage across a series circuit is also equal to the applied voltage, but consists of the sum of two or more individual voltage drops. This statement can be proven by an examination of the circuit shown in figure 9.40.

In this circuit a source potential (E_T) of 30 volts is impressed across a series circuit consisting of two 6 ohm resistors. The total resistance of the circuit is equal to the sum of the two individual resistances, or 12 ohms. Using Ohm's law, the circuit current may be calculated as follows:

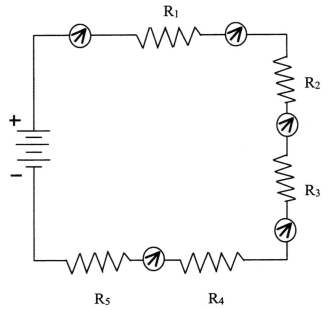

Figure 9.39.   Current in a series circuit.

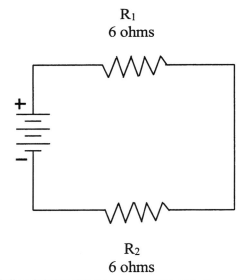

Figure 9.40.   Calculating total resistance in a series circuit.

$$I = \frac{E_T}{R_T}$$

$$I = \frac{30}{12}$$

$$I = 2.5 \text{ amperes}$$

Knowing the value of the resistors to be 6 ohms each, and the current through the resistors to be 2.5 amperes, the voltage drops across the resistors can be calculated. The voltage (E_1) across R_1 is therefore:

$$E_1 = IR_1$$
$$E_1 = 2.5 \text{ amps} \times 6 \text{ ohms}$$
$$E_1 = 15 \text{ volts}$$

Because R_2 is the same ohmic value as R_1 and carries the same current, the voltage drop across R_2 is also equal to 15 volts. Adding these two 15-volt drops together gives a total drop of 30 volts, exactly equal to the applied voltage. For a series circuit then:

$$E_T = E_1 + E_2 + E_3 \ldots E_n \qquad (9.11)$$

where

E_T = total voltage, V
E_1 = voltage across resistance R_1, V
E_2 = voltage across resistance R_2, V
E_3 = voltage across resistance R_3, V

Example 9.16
Problem

A series circuit consists of three resistors having values of 10 ohms, 20 ohms, and 40 ohms respectively. Find the applied voltage if the current through the 20-ohm resistor is 2.5 amperes.

Solution

To solve this problem, a circuit diagram is first drawn and labeled as shown in figure 9.41.

Given:

R_1 = 10 ohms
R_2 = 20 ohms
R_3 = 40 ohms
I = 2.5 amps

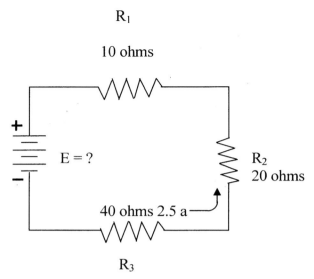

R_1

10 ohms

$E = ?$

R_2
20 ohms

40 ohms 2.5 a

R_3

Figure 9.41.   Solving for applied voltage in a series circuit.

Because the circuit involved is a series circuit, the same 2.5 amperes of current flows through each resistor. Using Ohm's law, the voltage drops across each of the three resistors can be calculated and are:

$E_1 = 25$ volts
$E_2 = 50$ volts
$E_3 = 100$ volts

Once the individual drops are known they can be added to find the total or applied voltage using equation 9.11:

$$E_T = E_1 + E_2 + E_3$$
$$E_T = 25 \text{ v} + 50 \text{ v} + 100 \text{ v}$$
$$E_T = 175 \text{ volts}$$

✔ **Key Point:** The total voltage (E_T) across a series circuit is equal to the sum of the voltages across each resistance of the circuit.

✔ **Key Point:** The voltage drops that occur in a series circuit are in direct proportion to the resistance across which they appear. This is the result of having the same current flow through each resistor. Thus the larger the resistor, the larger will be the voltage drop across it.

Power in Series Circuit

Each resistor in a series circuit consumes power. This power is dissipated in the form of heat. Since this power must come from the source, the total power must be equal in amount to the power consumed by the circuit resistances. In a series circuit the total power is equal to the sum of the powers dissipated by the individual resistors. Total power (P_T) is thus equal to:

$$P_T = P_1 + P_2 + P_3 \ldots P_n \tag{9.12}$$

where

P_T = total power, W
P_1 = power used in first part, W
P_2 = power used in second part, W
P_3 = power used in third part, W
P_n = power used in nth part, W

Example 9.17
Problem
A series circuit consists of three resistors having values of 5 ohms, 15 ohms, and 20 ohms. Find the total power dissipation when 120 volts is applied to the circuit (see figure 9.42).

Solution
Given:

R_1 = 5 ohms
R_2 = 15 ohms
R_3 = 20 ohms
E = 120 volts

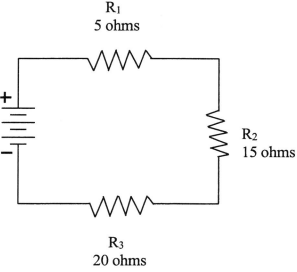

Figure 9.42. Solving for total power in a series circuit.

The total resistance is found first.

$$R_T = R_1 + R_2 + R_3$$
$$R_T = 5 + 15 + 20$$
$$R_T = 40 \text{ ohms}$$

Using total resistance and the applied voltage, the circuit current is calculated.

$$I = \frac{E_T}{R_T}$$

$$I = \frac{120}{40}$$

$$I = 3 \text{ amps}$$

Using the power formula, the individual power dissipations can be calculated.

For resistor R_1:

$$P_1 = I^2R_1$$
$$P_1 = (3)^25$$
$$P_1 = 45 \text{ watts}$$

For R_2:

$$P_2 = I^2R_2$$
$$P_2 = (3)^215$$
$$P_2 = 135 \text{ watts}$$

For R_3:

$$P_3 = I^2R_3$$
$$P_3 = (3)^220$$
$$P_3 = 180 \text{ watts}$$

To obtain total power:

$$P_T = P_1 + P_2 + P_3$$
$$P_T = 45 + 135 + 180$$
$$P_T = 360 \text{ watts}$$

To check the answer the total power delivered by the source can be calculated:

$$P = E \times I$$
$$P = 3\,a \times 120\,v$$
$$P = 360\ watts$$

Thus the total power is equal to the sum of the individual power dissipations.

✔ **Key Point:** We found that Ohm's law can be used for total values in a series circuit as well as for individual parts of the circuit. Similarly, the formula for power may be used for total values.

$$P_T = IE_T \tag{9.13}$$

Summary of the Rules for Series D.C. Circuits

To this point we have covered many of the important factors governing the operation of basic series circuits. In essence, what we have really done is to lay a strong foundation to build upon in preparation for more advanced circuit theory that follows. A summary of the important factors governing the operation of a series circuit are listed as follows:

1. The same current flows through each part of a series circuit.
2. The total resistance of a series circuit is equal to the sum of the individual resistances.
3. The total voltage across a series circuit is equal to the sum of the individual voltage drops.
4. The voltage drop across a resistor in a series circuit is proportional to the size of the resistor.
5. The total power dissipated in a series circuit is equal to the sum of the individual dissipations.

Parallel D.C. Circuits

The principles we applied to solving simple series circuit calculations for determining the reactions of such quantities as voltage, current, and resistance also can be used in parallel and series-parallel circuits.

A *parallel circuit* is defined as one having two or more components connected across the same voltage source (see figure 9.43). Recall that in a series circuit there is only one path for current flow. As additional loads (resistors, etc.) are added to the circuit, the total resistance increases and the total current decreases. This is not the case in a parallel circuit. In a parallel circuit, each load (or branch) is connected directly across the voltage source. In figure 9.43, commencing at the voltage source (E_b) and tracing counterclockwise around the circuit, two complete and separate paths can be identified in which current can flow. One path is traced from the source

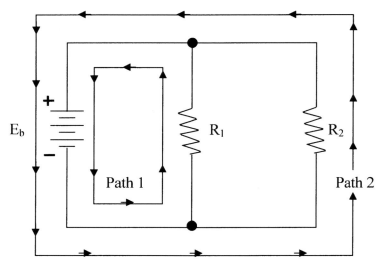

Figure 9.43.   Basic parallel circuit.

through resistance R_1 and back to the source; the other, from the source through resistance R_2 and back to the source.

Voltage in Parallel Circuits

Recall that in a series circuit the source voltage divides proportionately across each resistor in the circuit. In a parallel circuit (see figure 9.44), the same voltage is present across all the resistors of a parallel group. This voltage is equal to the applied voltage (E_b) and can be expressed in equation form as:

$$E_b = E_{R1} = E_{R2} = E_{Rn} \qquad (9.14)$$

We can verify equation 9.14 by taking voltage measurements across the resistors of a parallel circuit, as illustrated in figure 9.44. Notice that each voltmeter indicates the same amount of voltage—that is, the voltage across each resistor is the same as the applied voltage.

✔ **Key Point:** In a parallel circuit the voltage remains the same throughout the circuit.

Example 9.18
Problem
Assume that the current through a resistor of a parallel circuit is known to be 4.0 milliamperes (ma) and the value of the resistor is 40,000 ohms. Determine the potential (voltage) across the resistor. The circuit is shown in figure 9.45.

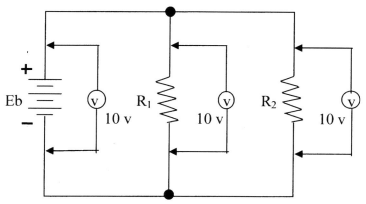

Figure 9.44.   Voltage comparison in a parallel circuit.

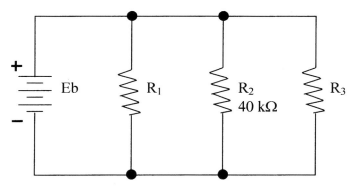

Figure 9.45.   (For example 9.18.)

Solution
Given:

$R_2 = 40 \text{ k}\Omega$
$I_{R2} = 4.0 \text{ ma}$

Find:

$E_{R2} = ?$
$E_b = ?$

Select the proper equation:

$$E = IR$$

Substitute known values:

$$E_{R2} = I_{R2} \times R_2$$
$$E_{R2} = 4.0 \text{ ma} \times 40,000 \text{ ohms}$$
[use power of tens]
$$E_{R2} = (4.0 \times 10^{-3}) \times (40 \times 10^3)$$
$$E_{R2} = 4.0 \times 40$$

Resultant:

$$E_{R2} = 160 \text{ v}$$

Therefore:

$$E_b = 160 \text{ v}$$

Current in Parallel Circuits

In a series circuit, a single current flows. Its value is determined in part by the total resistance of the circuit. However, the source current in a parallel circuit divides among the available paths in relation to the value of the resistors in the circuit. Ohm's law remains unchanged. For a given voltage, current varies inversely with resistance.

The behavior of current in a parallel circuit is best illustrated by example. The example we use is figure 9.46. The resistors R_1, R_2, and R_3 are in parallel with each other and with the battery. Each parallel path is then a branch with its own individual current. When the total current I_T leaves the voltage source E, part I_1 of the current I_T will flow through R_1, part I_2 will flow through R_2, and the remainder, I_3, through R_3. The branch current I_1, I_2, and I_3 can be different. However, if a voltmeter (used for measuring the voltage of a circuit) is connected across R_1, R_2, and R_3, the respective voltages, E_1, E_2, and E_3, will be equal. Therefore,

$$E = E_1 = E_2 = E_3 \tag{9.15}$$

The total current I_T is equal to the sum of all branch currents.

$$I_T = I_1 + I_2 + I_3 \tag{9.16}$$

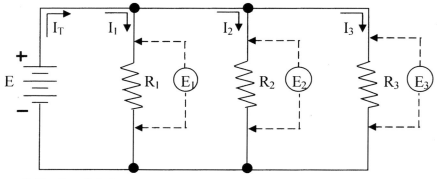

Figure 9.46. Parallel circuit.

This formula applies for any number of parallel branches, whether the resistances are equal or unequal.

By Ohm's law, each branch current equals the applied voltage divided by the resistance between the two points where the voltage is applied. Hence (see figure 9.46) for each branch we have the following equations:

$$\text{Branch 1:} \quad I_1 = \frac{E_1}{R_1} = \frac{E}{R_1}$$

$$\text{Branch 2:} \quad I_2 = \frac{E_2}{R_2} = \frac{E}{R_2}$$

$$\text{Branch 3:} \quad I_c = \frac{E_3}{R_3} = \frac{V}{R_3}$$

With the same applied voltage, any branch that has less resistance allows more current through it than a branch with higher resistance.

Example 9.19
Problem
Two resistors each drawing 2 amps and a third resistor drawing 1 amp are connected in parallel across a 100 V line (see figure 9.47). What is the total current?

Parallel Resistance

Unlike series circuits where total resistance (R_T) is the sum of the individual resistances, in a parallel circuit the total resistance is *not* the sum of the individual resistances.

In a parallel circuit we can use Ohm's law to find total resistance. We use the equation

$$R = \frac{E}{I}$$

or

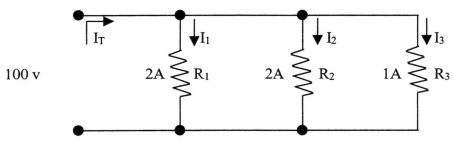

Figure 9.47.   (For example 9.19.)

$$R_T = \frac{E_s}{I_t}$$

R_T is the total resistance of all the parallel branches across the voltage source E_s, and I_T is the sum of all the branch currents.

Example 9.20
Problem
What is the total resistance of the circuit shown in figure 9.48?

Given:

$E_s = 120$ V
$I_T = 26$ A

Solution
In figure 9.48 the line voltage is 120 V and the total line current is 26 A. Therefore,

$$R_T = \frac{E}{I_T} \text{ eq } \frac{120}{26} = 4.62 \text{ ohms}$$

✔ **Important Point:** Notice that R_T is smaller than any of the three resistances in figure 9.48. This fact may surprise you—it may seem strange that the total circuit resistance is *less* than that of the smallest resistor (R_3—12 ohms). However, if we refer back to the water analogy we have used previously, it makes sense. Consider water pressure and water pipes—and assume there is some way to keep the water pressure constant. A small pipe offers more resistance to flow of water than a larger pipe; but if we add another pipe in parallel, one of even smaller diameter, the total resistance to water flow is *decreased*. In an electrical circuit, even a larger resistor in another parallel branch provides an *additional path for current flow*, so the *total* resistance is less. Remember, if we add one more branch to a parallel circuit, the total resistance decreases and the total current increases.

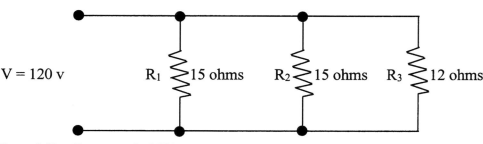

Figure 9.48.   (For example 9.20.)

Back to example 9.20 and figure 9.48: What we essentially demonstrated in working this particular problem is that the total load connected to the 120 V line is the same as the single equivalent resistance of 4.62 Ω connected across the line. (It is probably more accurate to call this total resistance the "equivalent resistance," but by convention R_T, or total resistance, is used—but they are often used interchangeably, too.)

The "equivalent resistance" is illustrated in the equivalent circuit shown in figure 9.49.

There are other methods used to determine the equivalent resistance of parallel circuits. The most appropriate method for a particular circuit depends on the number and value of the resistors. For example, consider the parallel circuit shown in figure 9.50.

For this circuit, the following simple equation is used:

$$R_{eq} = \frac{R}{N} \tag{9.17}$$

where

R_{eq} = equivalent parallel resistance
R = ohmic value of one resistor
N = number of resistors

Thus,

$$R_{eq} = \frac{10 \text{ ohms}}{2}$$

$$R_{eq} = 5 \text{ ohms}$$

✔ **Key Point:** When two equal value resistors are connected in parallel, they present a total resistance equivalent to a single resistor of one-half the value of either of the original resistors.

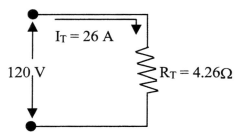

$I_T = 26$ A

120 V

$R_T = 4.26\Omega$

Figure 9.49.   Equivalent circuit to that of figure 9.48.

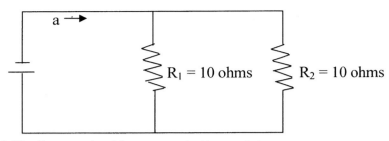

Figure 9.50.   Two equal resistors connected in parallel.

Example 9.21
Problem
Five 50-ohm resistors are connected in parallel. What is the equivalent circuit resistance?

Solution

$$R_{eq} = \frac{R}{N} = \frac{50}{5} = 10 \text{ ohms}$$

What about parallel circuits containing resistance of unequal value? How is equivalent resistance determined? Example 9.22 demonstrates how this is accomplished.

Example 9.22
Problem
Refer to figure 9.51.

Solution
Given:

$R_1 = 3\Omega$
$R_2 = 6\Omega$
$E_a = 30$ V

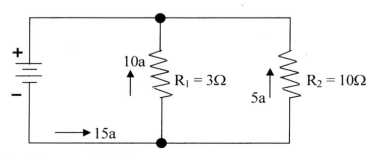

Figure 9.51.   (For example 9.22.)

Known:

I_1 = 10 amps
I_2 = 5 amps
I_t = 15 amps

Determine:

R_{eq} = ?

$$R_{eq} = \frac{E_a}{I_t}$$

$$R_{eq} = \frac{30}{15} = 2 \text{ ohms}$$

✔ **Key Point:** In example 9.22 the equivalent resistance of 2 ohms is less than the value of either branch resistor. Remember, in parallel circuits the equivalent resistance will always be smaller than the resistance of any branch.

Reciprocal Method

When circuits are encountered in which resistors of unequal value are connected in parallel, the equivalent resistance may be computed by using the *reciprocal method*.

✔ **Note:** A *reciprocal* is an inverted fraction; the reciprocal of the fraction 3/4, for example is 4/3. We consider a whole number to be a fraction with 1 as the denominator, so the reciprocal of a whole number is that number divided into 1. For example, the reciprocal of R_T is $1/R_T$.

The equivalent resistance in parallel is given by the formula

$$\frac{1}{R_T} = \frac{1}{R_1} + \frac{1}{R_2} + \frac{1}{R_3} + \ldots + \frac{1}{R_n} \qquad (9.18)$$

where

R_T is the total resistance in parallel
R_1, R_2, R_3, and R_n are the branch resistances

Example 9.23
Problem
Find the total resistance of a 2-ohm, a 4-ohm, and an 8-ohm resistor in parallel (figure 9.52).

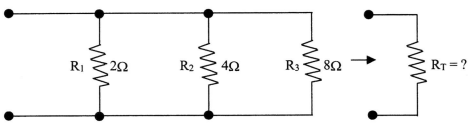

Figure 9.52.   (For example 9.23.)

Solution
Write the formula for three resistors in parallel.

$$\frac{1}{R_T} = \frac{1}{R_1} + \frac{1}{R_2} + \frac{1}{R_3}$$

Substitute the resistance values.

$$\frac{1}{R_T} = \frac{1}{2} + \frac{1}{4} + \frac{1}{8}$$

Add fractions.

$$\frac{1}{R_T} = \frac{4}{2} + \frac{2}{8} + \frac{1}{8} = \frac{7}{8}$$

Invert both sides of the equation to solve for R_T.

$$R_T = \frac{8}{7} = 1.14\ \Omega$$

✔ **Note:** When resistances are connected in parallel, the total resistance is always *less* than the smallest resistance of any single branch.

Product-over-the-Sum Method

When any two unequal resistors are in parallel, it is often easier to calculate the total resistance by multiplying the two resistances and then dividing the product by the sum of the resistances.

$$R_T = \frac{R_1 \times R_2}{R_1 + R_2} \tag{9.19}$$

where

R_T is the total resistance in parallel
R_1 and R_2 are the two resistors in parallel

Example 9.24
Problem
What is the equivalent resistance of a 20-ohm and a 30-ohm resistor connected in parallel?

Solution
Given:

$R_1 = 20$
$R_2 = 30$

$$R_T = \frac{R_1 \times R_2}{R_1 + R_2}$$

$$R_T = \frac{20 \times 30}{20 + 30}$$

$$R_T = 12 \text{ ohms}$$

Power in Parallel Circuits

As in the series circuit, the total *power* consumed in a parallel circuit is equal to the sum of the power consumed in the individual resistors.

✔ **Note:** Because power dissipation in resistors consists of a heat loss, power dissipations are additive regardless of how the resistors are connected in the circuit.

$$P_T = P_1 + P_2 + P_3 + \ldots + P_n \tag{9.20}$$

where

P_T is the total power
P_1, P_2, P_3, and P_n are the branch powers

Total power can also be calculated by the equation

$$P_T = EI_T \tag{9.21}$$

where

P_T is the total power
E is the voltage source across all parallel branches
I_T is the total current

The power dissipated in each branch is equal to EI and equal to V^2/R.

✔ **Note:** In both parallel and series arrangements the sum of the individual values of power dissipated in the circuit equals the total power generated by the source. The circuit arrangements cannot change the fact that all power in the circuit comes from the source.

Example 9.25
Problem
Find the total power consumed by the circuit in figure 9.53.

Solution

$$P_{R1} = E_b \times I_{R1}$$
$$P_{R1} = 50 \times 5$$
$$P_{R1} = 250 \text{ watts}$$

$$P_{R2} = E_b \times I_{R2}$$
$$P_{R2} = 50 \times 2$$
$$P_{R2} = 100 \text{ watts}$$

$$P_{R3} = E_b \times I_{R3}$$
$$P_{R3} = 50 \times 1$$
$$P_{R3} = 50 \text{ watts}$$

$$P_T = P_1 + P_2 + P_3$$
$$P_T = 250 + 100 + 50$$
$$P_T = 400 \text{ watts}$$

✔ **Note:** The power dissipated in the branch circuits in figure 9.53 is determined in the same manner as the power dissipated by individual resistors in a series circuit.

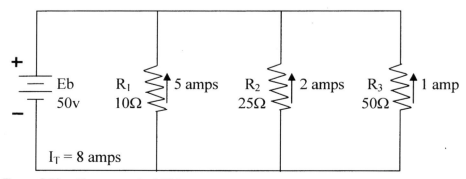

Figure 9.53.   (For example 9.25.)

The total power (P_T) is then obtained by summing up the powers dissipated in the branch resistors using equation 9.20.

Because, in the example shown in figure 9.53, the total current is known, we could determine the total power by the following method:

$$P_T = E_b \times I_T$$
$$P_T = 50 \text{ v} \times 8a$$
$$P_T = 400 \text{ watts}$$

Rules for Solving Parallel D.C. Circuits

Problems involving the determination of resistance, voltage, current, and power in a parallel circuit are solved as simply as in a series circuit. The procedure is basically the same: (1) draw a circuit diagram; (2) state the values given and the values to be found; (3) state the applicable equations; and (4) substitute the given values and solve for the unknown.

Along with following the problem-solving procedure above, it is also important to remember and apply the rules for solving parallel d.c. circuits. These rules are:

1. The same voltage exists across each branch of a parallel circuit and is equal to the source voltage.
2. The current through a branch of a parallel network is inversely proportional to the amount of resistance of the branch.
3. The total current of a parallel circuit is equal to the sum of the currents of the individual branches of the circuit.
4. The total resistance of a parallel circuit is equal to the reciprocal of the sum of the reciprocals of the individual resistances of the circuit.
5. The total power consumed in a parallel circuit is equal to the sum of the power consumption of the individual resistances.

Series-Parallel Circuits

To this point we have discussed series and parallel d.c. circuits. However, you will seldom encounter a circuit that consists solely of either type of circuit. Most circuits consist of both series and parallel elements. A circuit of this type is referred to as a *series-parallel circuit* (see figure 9.54) or as a *combination circuit*. The solution of a series-parallel circuit is simply a matter of application of the laws and rules discussed prior to this point.

A.C. Theory

Because voltage is induced in a conductor when lines of force are cut, the amount of the induced emf depends on the number of lines cut in a unit time. To induce an emf

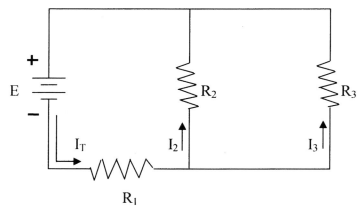

Figure 9.54.   A series-parallel circuit.

of 1 volt, a conductor must cut 100,000,000 lines of force per second. To obtain this great number of "cuttings," the conductor is formed into a loop and rotated on an axis at great speed (see figure 9.55). The two sides of the loop become individual conductors in series, each side of the loop cutting lines of force and inducing twice the voltage that a single conductor would induce. In commercial generators, the number of "cuttings" and the resulting emf are increased by: (1) increasing the number of lines of force by using more magnets or stronger electromagnets; (2) using more conductors or loops; and (3) rotating the loops faster.

How an a.c. generator operates to produce an a.c. voltage and current is a basic concept today, taught in elementary and middle school science classes. Of course, we accept technological advances as commonplace today. We surf the Internet, Google,

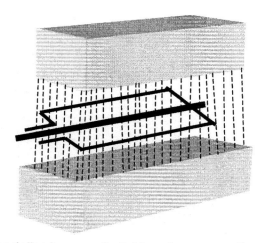

Figure 9.55.   Loop rotating in magnetic field produces a.c. voltage.

or Yahoo for answers, watch cable television, use our cell phones, and take space flight as a given—and we consider producing the electricity that makes all these technologies possible as our right. These technologies are bottom-shelf to us today. We have them available to us so we simply use them. This point of view surely was not held initially, especially by those who broke ground in developing technology and electricity.

In groundbreaking years of electric technology development, the geniuses of the science of electricity (including George Simon Ohm) performed their technological breakthroughs in faltering steps. We tend to forget that those first faltering steps of scientific achievement in the field of electricity were performed with crude, and for the most part, homemade apparatus. Indeed, the innovators of electricity had to fabricate nearly all the laboratory equipment used in their experiments. At the time, the only convenient source of electrical energy available to these early scientists was the voltaic cell, invented some years earlier. Because of the fact that cells and batteries were the only sources of power available, some of the early electrical devices were designed to operate from *direct current.*

Thus, initially, direct current was used extensively. However, when the use of electricity became widespread, certain disadvantages in the use of direct current became apparent. In a direct-current system, the supply voltage must be generated at the level required by the load. To operate a 240 V lamp for example, the generator must deliver 240 V. A 120 V lamp could not be operated from this generator by any convenient means. A resistor could be placed in series with the 120 V lamp to drop the extra 120 volts, but the resistor would waste an amount of power equal to that consumed by the lamp.

Another disadvantage of direct-current systems is the large amount of power lost due to the resistance of the transmission wires used to carry current from the generating station to the consumer. This loss could be greatly reduced by operating the transmission line at very high voltage and low current. This is not a practical solution in a d.c. system, however, since the load would also have to operate at high voltage. As a result of the difficulties encountered with direct current, practically all modern power distribution systems use *alternating current (a.c.)*, including water/wastewater treatment plants.

Unlike d.c. voltage, a.c. voltage can be stepped up or down by a device called a transformer. Transformers permit the transmission lines to be operated at high voltage and low current for maximum efficiency. Then at the consumer end the voltage is stepped down to whatever value the load requires by using a transformer. Due to its inherent advantages and versatility, alternating current has replaced direct current in all but a few commercial power distribution systems.

BASIC A.C. GENERATOR

In figure 9.55, an a.c. voltage and current is produced when a conductor loop rotates through a magnetic field and cuts lines of force to generate an induced a.c. voltage across its terminals. This describes the basic principle of operation of an alternating current generator, or alternator. An alternator converts mechanical energy into electrical energy. It does this by utilizing the principle of electromagnetic induction. The basic components of an alternator are an armature (about which many turns of conductor are wound) that rotates in a magnetic field, and some means of delivering the resulting alternating current to an external circuit.

CYCLE

A.C. voltage is one that continually changes in magnitude and periodically reverses in polarity (see figure 9.56). The zero axis is a horizontal line across the center. The vertical variations on the voltage wave show the changes in magnitude. The voltages above the horizontal axis have positive (+) polarity, while voltages below the horizontal axis have negative (−) polarity.

Figure 9.57 shows a suspended loop of wire (conductor or armature) being rotated (moved) in a counterclockwise direction through the magnetic field between the poles of a permanent magnet. For ease of explanation, the loop has been divided into a thick and thin half. Notice that in part (A), the thick half is moving along (parallel to) the lines of force. Consequently, it is cutting none of these lines. The same is true of the thin half, moving in the opposite direction. Since the conductors are not cutting any lines of force, no emf is induced. As the loop rotates toward the position shown in part (B), it cuts more and more lines of force per second because it is cutting more directly across the field (lines of force) as it approaches the position shown in (B). At position (B) the induced voltage is greatest because the conductor is cutting directly across the field.

As the loop continues to be rotated toward the position shown in part (C), it cuts fewer and fewer lines of force per second. The induced voltage decreases from its peak value. Eventually, the loop is once again moving in a plane parallel to the magnetic field, and no voltage (zero voltage) is induced. The loop has now been rotated through half a circle (one alternation, or 180°). The sine curve shown in the lower part of figure 9.57 shows the induced voltage at every instant of rotation of the loop. Notice

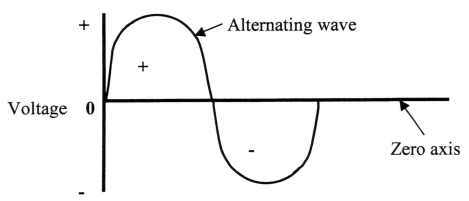

Figure 9.56. A.C. voltage waveform.

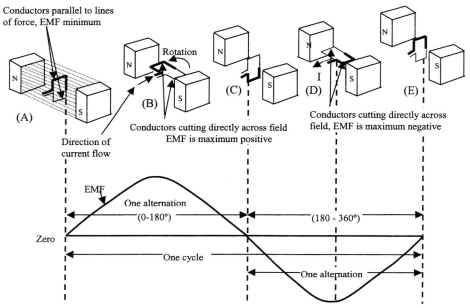

Figure 9.57.   Basic a.c. generator.

that this curve contains 360°, or two alternations. Two alternations represent one complete circle of rotation.

✔ **Important Point:** Two complete alternations in a period of time is called a *cycle*.

In figure 9.57, if the loop is rotated at a steady rate, and if the strength of the magnetic field is uniform, the number of cycles per second (cps), or hertz, and the voltage will remain at fixed values. Continuous rotation will produce a series of sine-wave voltage cycles, or, in other words, an a.c. voltage. In this way mechanical energy is converted into electrical energy.

FREQUENCY, PERIOD, AND WAVELENGTH

A.C. voltage is one that continually changes in magnitude and periodically reverses in polarity (see figure 9.58). The zero axis is a horizontal line across the center. The vertical variations on the voltage wave show the changes in magnitude. The voltages above the horizontal axis have positive (+) polarity, while voltages below the horizontal axis have negative (−) polarity.

The *frequency* of an alternating voltage or current is the number of complete cycles occurring in each second of time. It is indicated by the symbol f and is expressed in hertz (Hz). One cycle per second equals one hertz. Thus 60 cycles per second (cps) equals 60 Hz. A frequency of 2 Hz (figure 9.58 [A]) is twice the frequency of 1 Hz (figure 9.58 [B]).

The amount of time for the completion of 1 cycle is the *period*. It is indicated by the symbol T for time and is expressed in seconds (s). Frequency and period are reciprocals of each other.

$$f = \frac{1}{T} \qquad\qquad (9.22)$$

$$T = \frac{1}{f} \qquad\qquad (9.23)$$

✔ **Important Point**: The higher the frequency, the shorter the period.

The angle of 360° represents the time for 1 cycle, or the period T. So we can show the horizontal axis of the sine wave in units of either electrical degrees or seconds (see figure 9.59).

The *wavelength* is the length of one complete wave or cycle. It depends upon the frequency of the periodic variation and its velocity of transmission. It is indicated by the symbol λ (Greek letter lambda). Expressed as a formula:

$$\lambda = \frac{\text{velocity}}{\text{frequency}} \qquad\qquad (9.24)$$

VALUES OF A.C. VOLTAGE AND CURRENT

Because a.c. sine wave voltage or current has many instantaneous values throughout the cycle, it is convenient to specify magnitudes for comparing one wave with another. The peak, average, or root-mean-square (rms) value can be specified (see figure 9.60). These values apply to current or voltage.

PEAK AMPLITUDE

One of the most frequently measured characteristics of a sine wave is its amplitude. Unlike d.c. measurement, the amount of alternating current or voltage present in a

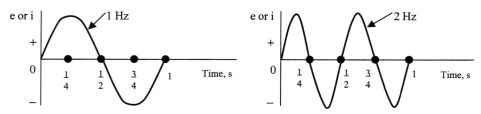

Figure 9.58. Comparison of frequencies.

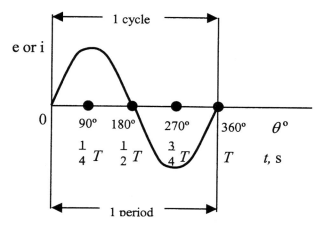

Figure 9.59.   Relationship between electrical degrees and time.

circuit can be measured in various ways. In one method of measurement, the maximum amplitude of either the positive or the negative alternation is measured. The value of current or voltage obtained is called the *peak voltage* or the *peak current*. To measure the peak value of current or voltage, an oscilloscope must be used. The peak value is illustrated in figure 9.60.

PEAK-TO-PEAK AMPLITUDE

A second method of indicating the amplitude of a sine wave consists of determining the total voltage or current between the positive and negative peaks. This value of

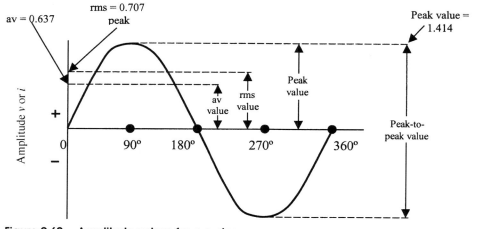

Figure 9.60.   Amplitude values for a.c. sine wave.

current or voltage is called the *peak-to-peak value* (see figure 9.60). Because both alternations of a pure sine wave are identical, the peak-to-peak value is twice the peak value. Peak-to-peak voltage is usually measured with an oscilloscope, although some voltmeters have a special scale calibrated in peak-to-peak volts.

INSTANTANEOUS AMPLITUDE

The *instantaneous value* of a sine wave of voltage for any angle of rotation is expressed by the formula:

$$e = E_m \times \sin \theta \qquad (9.25)$$

where

e = the instantaneous voltage
E_m = the maximum or peak voltage
$\sin \theta$ = the sine of angle at which e is desired

Similarly the equation for the instantaneous value of a sine wave of current would be:

$$i = I_m \times \sin \theta \qquad (9.26)$$

where

i = the instantaneous current
I_m = the maximum or peak current
$\sin \theta$ = the sine of the angle at which i is desired

✔ Note: The instantaneous value of voltage constantly changes as the armature of an alternator moves through a complete rotation. Because current varies directly with voltage, according to Ohm's law, the instantaneous changes in current also result in a sine wave whose positive and negative peaks and intermediate values can be plotted exactly as we plotted the voltage sine wave. However, instantaneous values are not useful in solving most a.c. problems, so an *effective* value is used.

EFFECTIVE OR RMS VALUE

The *effective value* of a.c. voltage or current of a sine waveform is defined in terms of an equivalent heating effect of a direct current. Heating effect is independent of the direction of current flow.

✔ **Important Point:** Because all instantaneous values of induced voltage are some-where between zero and E_M (the maximum voltage or peak voltage), the effective value of a sine wave voltage or current must be greater than zero and less than E_M.

The alternating current of sine waveform having a maximum value of 14.14 amps produces the same amount of heat in a circuit having a resistance of 1 ohm as a direct current of 10 amps. Because this is true, we can work out a constant value for convert-ing any peak value to a corresponding effective value. This constant is represented by X in the simple equation below. Solve for X to three decimal places.

$$14.14X = 10$$
$$X = 0.707$$

The effective value is also called the *root-mean-square (rms)* value, because it is the square root of the average of the squared values between zero and maximum. The effective value of a.c. current is stated in terms of an equivalent d.c. current. The phenomenon used as the standard comparison is the heating effect of the current.

✔ **Important Point:** Anytime an a.c. voltage or current is stated without any quali-fications, it is assumed to be an effective value.

In many instances it is necessary to convert from effective to peak or vice versa using a standard equation. Figure 9.60 shows that the peak value of a sine wave is 1.414 times the effective value; therefore the equations we use are:

$$E_m = E \times 1.414 \tag{9.27}$$

where

E_m = maximum or peak voltage
E = effective or RMS voltage

and

$$I_m = I \times 1.414 \tag{9.28}$$

where

I_m = maximum or peak current
I = effective or RMS current

Upon occasion it is necessary to convert a peak value of current or voltage to an effective value. This is accomplished by using the following equations:

$$E = E_m \times 0.707 \tag{9.29}$$

where

E = effective voltage
E_m = the maximum or peak voltage

and

$$I = I_m \times 0.707 \tag{9.30}$$

where

I = the effective current
I_m = the maximum or peak current

AVERAGE VALUE

Because the positive alternation is identical to the negative alternation, the *average value* of a complete cycle of a sine wave is zero. In certain types of circuits however, it is necessary to compute the average value of one alternation. Figure 9.60 shows that the average value of a sine wave is 0.637 × peak value and therefore:

$$\text{Average Value} = 0.637 \times \text{peak value} \tag{9.31}$$

or

$$E_{avg} = E_m \times 0.637$$

where

E_{avg} = the average voltage of one alternation
E_m = the maximum or peak voltage

Similarly:

$$I_{avg} = I_m \times 0.637 \tag{9.32}$$

where

I_{avg} = the average current in one alternation
I_m = the maximum or peak current

Table 9.5 lists the various values of sine wave amplitude used to multiply in the conversion of a.c. sine wave voltage and current.

Table 9.5   A.C. sine wave conversion

Multiply the value	By	To get the value
Peak	2	Peak-to-peak
Peak-to-peak	0.5	Peak
Peak	0.637	Average
Average	1.637	Peak
Peak	0.707	RMS (effective)
RMS (effective)	1.414	Peak
Average	1.110	RMS (effective)
RMS (effective)	0.901	Average

RESISTANCE IN A.C. CIRCUITS

If a sine wave of voltage is applied to a resistance, the resulting current will also be a sine wave. This follows Ohm's law, which states that the current is directly proportional to the applied voltage. Figure 9.61 shows a sine wave of voltage and the resulting sine wave of current superimposed on the same time axis. Notice that as the voltage increases in a positive direction the current increases along with it. When the voltage reverses direction, the current reverses direction. At all times the voltage and current pass through the same relative parts of their respective cycles at the same time. When two waves, such as those shown in figure 9.61, are precisely in step with one another they are said to be *in phase*. To be in phase, the two waves must go through their maximum and minimum points at the same time and in the same direction.

In some circuits, several sine waves can be in phase with each other. Thus, it is possible to have two or more voltage drops in phase with each other and also in phase with the circuit current.

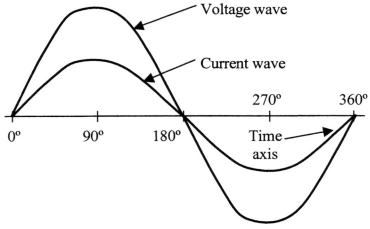

Figure 9.61.   Voltage and current waves in phase.

✔ **Note:** It is important to remember that Ohm's law for d.c. circuits is applicable to a.c. circuits *with resistance only*.

Voltage waves are not always in phase. For example, figure 9.62 shows a voltage wave E_1 considered to start at 0° (time 1). As voltage wave E_1 reaches its positive peak, a second voltage wave (E_2) starts to rise (time 2). Because these waves do not go through their maximum and minimum points at the same instant of time, a *phase difference* exists between the two waves. The two waves are said to be out of phase. For the two waves in figure 9.62, this phase difference is 90°.

PHASE RELATIONSHIPS

In the preceding section we discussed the important concepts of *in phase* and *phase difference*. Another important phase concept is *phase angle*. The phase angle between two waveforms of the same frequency is the angular difference at a given instant of time. As an example, the phase angle between waves B and A (see figure 9.63) is 90°. Take the instant of time at 90°. The horizontal axis is shown in angular units of time. Wave B starts at maximum value and reduces to zero value at 90°, while wave A starts at zero and increases to maximum value at 90°. Wave B reaches its maximum value 90° ahead of wave A, so wave B *leads* wave A by 90° (and wave A *lags* wave B by 90°). This 90° phase angle between waves B and A is maintained throughout the complete cycle and all successive cycles. At any instant of time, wave B has the value that wave

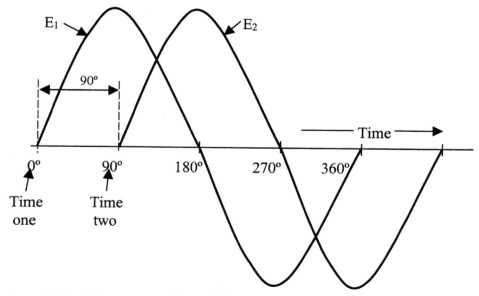

Figure 9.62.   Voltage waves 90° out of phase.

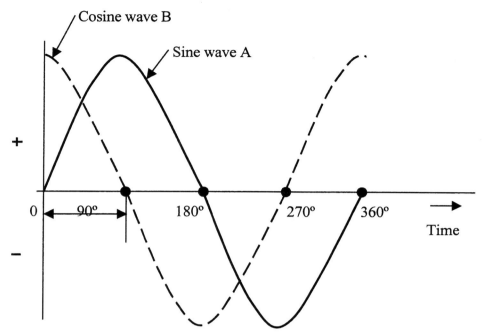

Figure 9.63.   Wave B leads wave A by a phase angle of 90°.

A will have 90° later. Wave B is a cosine wave because it is displaced 90° from wave A, which is a sine wave.

✔ **Important Point**: The amount by which one wave leads or lags another is measured in degrees.

To compare phase angles or phases of alternating voltages or currents, it is more convenient to use vector diagrams corresponding to the voltage and current waveforms. A *vector* is a straight line used to denote the magnitude and direction of a given quantity. Magnitude is denoted by the length of the line drawn to scale, and the direction is indicated by the arrow at one end of the line, together with the angle that the vector makes with a horizontal reference vector.

✔ **Note**: In electricity, since different directions really represent *time* expressed as a phase relationship, an electrical vector is called a *phasor*. In an a.c. circuit containing only resistance, the voltage and current occur at the *same time*, or are in phase. To indicate this condition by means of phasors, all that is necessary is to draw the phasors for the voltage and current in the same direction. The value of each is indicated by the *length* of the phasor.

A vector, or phasor, diagram is shown in figure 9.64 where vector V_B is vertical to show the phase angle of 90° with respect to vector V_A, which is the reference.

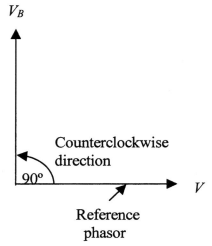

Figure 9.64.   Phasor diagram.

Because lead angles are shown in the counterclockwise direction from the reference vector, V_B leads V_A by 90°.

Inductance

To this point we have learned the following key points about magnetic fields:

- A field of force exists around a wire carrying a current.
- This field has the form of concentric circles around the wire, in planes perpendicular to the wire, and with the wire at the center of the circles.
- The strength of the field depends on the current. Large currents produce large fields; small currents produce small fields.
- When lines of force cut across a conductor, a voltage is induced in the conductor.

Moreover, to this point we have studied circuits that have been resistive (i.e., resistors presented the only opposition to current flow). Two other phenomena—inductance and capacitance—exist in d.c. circuits to some extent, but they are major players in a.c. circuits. Both inductance and capacitance present a kind of opposition to current flow that is called *reactance*.

INDUCTANCE: WHAT IS IT?

Inductance is the characteristic of an electrical circuit that makes itself evident by opposing the starting, stopping, or changing of current flow. A simple analogy can be used to explain inductance. We are all familiar with how difficult it is to push a heavy load (a cart full of heavy materials, etc.). It takes more work to start the load moving than it does to keep it moving. This is because the load possesses the property of inertia. *Inertia* is the characteristic of mass that opposes a change in velocity. There-fore, inertia can hinder us in some ways and help us in others. Inductance exhibits the same effect on current in an electric circuit as inertia does on velocity of a mechanical object. The effects of inductance are sometimes desirable and sometimes undesirable.

✔ **Important Point:** Simply put, inductance is the characteristic of an electrical conductor that opposes a *change* in current flow.

What does all this mean? Good question.

What it means is that since inductance is the property of an electric circuit that opposes any change in the current through that circuit, if the current increases, a self-induced voltage opposes this change and delays the increase. On the other hand, if the current decreases, a self-induced voltage tends to aid (or prolong) the current flow, delaying the decrease. Thus, current can neither increase nor decrease as fast in an inductive circuit as it can in a purely resistive circuit.

In a.c. circuits, this effect becomes very important because it affects the phase relationships between voltage and current. We learned that voltages (or currents) can be out of phase if they are induced in separate armatures of an alternator. In that case, the voltage and current generated by each armature were in phase. When inductance is a factor in a circuit, the voltage and current generated by the *same* armature are out of phase.

Unit of Inductance

The unit for measuring inductance, L, is the *henry* (named for the American physicist, Joseph Henry), abbreviated *h*. Figure 9.65 shows the schematic symbol for an induc-tor. An inductor has an inductance of 1 henry if an emf of 1 volt is induced in the inductor when the current through the inductor is changing at the rate of 1 ampere per second. The relation between the induced voltage, inductance, and rate of change of current with respect to time is stated mathematically as

$$E = L\frac{\Delta I}{\Delta t} \tag{9.33}$$

where

E = the induced emf in volts
L = the inductance in henrys
ΔI = is the change in amperes occurring in Δt seconds

Figure 9.65. Schematic symbol for an inductor.

✔ **Note:** Recall that the symbol Δ (delta) means "a change in . . .".

The henry is a large unit of inductance and is used with relatively large inductors. The unit employed with small inductors is the millihenry (mh). For still smaller inductors the unit of inductance is the microhenry (μh).

Self-Inductance

As previously explained, current flow in a conductor always produces a magnetic field surrounding, or linking with, the conductor. When the current changes, the magnetic field changes, and an emf is induced in the conductor. This emf is called a *self-induced emf* because it is induced in the conductor carrying the current.

✔ **Note:** Even a perfectly straight length of conductor has some inductance.

The direction of the induced emf has a definite relation to the direction in which the field that induces the emf varies. When the current in a circuit is increasing, the flux linking with the circuit is increasing. This flux cuts across the conductor and induces an emf in the conductor in such a direction to oppose the increase in current and flux. This emf is sometimes referred to as *counterelectromotive force* (cemf). The two terms are used synonymously throughout this chapter. Likewise, when the current is decreasing, an emf is induced in the opposite direction and opposes the decrease in current.

✔ **Important Point:** The effects just described are summarized by Lenz's law, which states that the induced emf in any circuit is always in a direction opposed to the effect that produced it.

Shaping a conductor so that the electromagnetic field around each portion of the conductor cuts across some other portion of the same conductor increases the inductance. This is shown in its simplest form in figure 9.66 (A). A loop of conductor is looped so that two portions of the conductor lie adjacent and parallel to one another. These portions are labeled Conductor 1 and Conductor 2. When the switch is closed, electron flow through the conductor establishes a typical concentric field around all portions of the conductor. The field is shown in a single plane (for simplicity) that is perpendicular to both conductors. Although the field originates simultaneously in both conductors, it is considered as originating in Conductor 1, and its effect on Conductor 2 will be noted. With increasing current, the field expands outward, cutting across a portion of Conductor 2. The resultant induced emf in Conductor 2 is shown by the dashed arrow. Note that it is in opposition to the battery current and voltage, according to Lenz's law.

In figure 9.66 (B), the same section of Conductor 2 is shown, but with the switch opened and the flux collapsing.

✔ **Important Point:** From figure 9.66, the important point to note is that the voltage of self-induction opposes both changes in current. It delays the initial

(A)

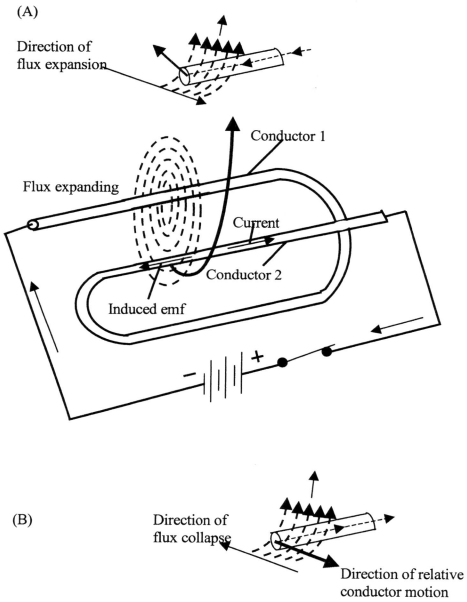

Direction of
flux expansion

Conductor 1

Flux expanding

Current

Conductor 2

Induced emf

+

−

(B)

Direction of
flux collapse

Direction of relative
conductor motion

Figure 9.66. Self-inductance.

buildup of current by opposing the battery voltage, and it delays the breakdown of current by exerting an induced voltage in the same direction that the battery voltage acted.

Four major factors affect the self-inductance of a conductor, or circuit:

1. **Number of turns**: Inductance depends on the number of wire turns. Wind more turns to increase inductance. Take turns off to decrease the inductance. Figure 9.67 compares the inductance of two coils made with different numbers of turns.
2. **Spacing between turns**: Inductance depends on the spacing between turns, or the inductor's length. Figure 9.68 shows two inductors with the same number of turns. The first inductor's turns have a wide spacing. The second inductor's turns are close together. The second coil, though shorter, has a larger inductance value because of its close spacing between turns.
3. **Coil diameter**: Coil diameter, or cross-sectional area, is highlighted in figure 9.69.

(A)   (B)

Figure 9.67.   (A) Few turns, low inductance. (B) More turns, higher inductance.

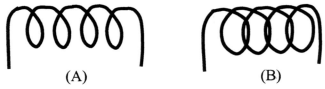

(A)   (B)

Figure 9.68.   (A) Wide spacing between turns, low inductance. (B) Close spacing between turns, higher inductance.

(A)   (B)

Figure 9.69.   (A) Small diameter, low inductance. (B) Larger diameter, higher inductance.

The larger-diameter inductor has more inductance. Both coils shown have the same number of turns, and the spacing between turns is the same. The first inductor has a small diameter and the second one has a larger diameter. The second inductor has more inductance than the first one.

4. **Type of core material**: Permeability, as pointed out earlier, is a measure of how easily a magnetic field goes through a material. Permeability also tells us how much stronger the magnetic field will be with the material inside the coil. Figure 9.70 shows three identical coils. One has an air core, one has a powdered-iron core in the center, and the other has a soft iron core. This figure illustrates the effects of core material on inductance. The inductance of a coil is affected by the magnitude of current when the core is a magnetic material. When the core is air, the inductance is independent of the current.

✔ **Key Point:** The inductance of a coil increases very rapidly as the number of turns is increased. It also increases as the coil is made shorter, the cross-sectional area is made larger, or the permeability of the core is increased.

Mutual Inductance

When the current in a conductor or coil changes, the varying flux can cut across any other conductor or coil located nearby, thus inducing voltages in both. A varying current in L_1, therefore, induces voltage across L_1 and across L_2 (see figure 9.71; see figure 9.72 for the schematic symbol for two coils with mutual inductance). When the induced voltage e_{L2} produces current in L_2, its varying magnetic field induces voltage in L_1. Hence, the two coils L_1 and L_2 have *mutual inductance* because current change in one coil can induce voltage in the other. The unit of mutual inductance is the henry, and the symbol is L_M. Two coils have an L_M of 1 H when a current change of 1 A/s in one coil induces 1 E in the other coil.

The factors affecting the mutual inductance of two adjacent coils is dependent upon:

- the physical dimensions of the two coils
- the number of turns in each coil
- the distance between the two coils
- the relative positions of the axes of the two coils
- the permeability of the cores

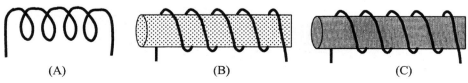

Figure 9.70. (A) Air core, low inductance. (B) Powdered iron core, higher inductance. (C) Soft iron core, highest inductance.

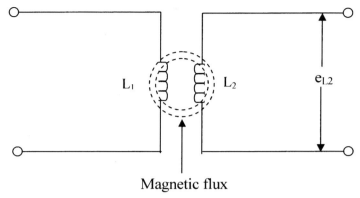

Magnetic flux

Figure 9.71.   Mutual inductance between L₁ and L₂.

✔ **Important Point:** The amount of mutual inductance depends on the relative position of the two coils. If the coils are separated a considerable distance, the amount of flux common to both coils is small and the mutual inductance is low. Conversely, if the coils are close together so that nearly all the flow of one coil links the turns of the other, mutual inductance is high. The mutual inductance can be increased greatly by mounting the coils on a common iron core.

Calculation of Total Inductance

If inductors in series are located far enough apart, or well shielded to make the effects of mutual inductance negligible, the total inductance is calculated in the same manner as for resistances in series—we merely add them:

$$L_T = L_1 + L_2 + L_3 \ldots \text{(etc.)} \tag{9.34}$$

Example 9.26

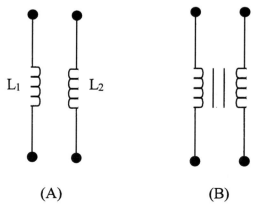

(A)   (B)

Figure 9.72.   (A) Schematic symbol for two coils (air core) with mutual inductance. (B) Two coils (iron core) with mutual inductance.

Problem

If a series circuit contains three inductors whose values are 40 µh, 50 µh, and 20 µh, what is the total inductance?

Solution

$$L_T = 40 \ \mu h + 50 \ \mu h + 20 \ \mu$$
$$= 110 \ \mu h$$

In a parallel circuit containing inductors (without mutual inductance), the total inductance is calculated in the same manner as for resistances in parallel:

$$\frac{1}{L_T} = \frac{1}{L_1} + \frac{1}{L_2} + \frac{1}{L_3} + \ldots \text{(etc.)} \tag{9.35}$$

Example 9.27
Problem

A circuit contains three totally shielded inductors in parallel. The values of the three inductances are: 4 mh, 5 mh, and 10 mh. What is the total inductance?

Solution

$$\frac{1}{L_T} = \frac{1}{4} + \frac{1}{5} + \frac{1}{10}$$

$$= 0.25 + 0.2 + 0.1$$

$$= 0.55$$

$$L_T = \frac{1}{0.55}$$

$$= 1.8 \ mh$$

Capacitance

No matter how complex the electrical circuit, it is composed of no more than three basic electrical properties: resistance, inductance, and capacitance. Accordingly, understanding each of these three basic properties is a necessary step toward understanding electrical equipment. The last of the basic three, capacitance, is covered here.

   In the previous section, we learned that inductance opposes any change in current. *Capacitance* is the property of an electric circuit that opposes any change of *voltage* in a circuit. That is, if applied voltage is increased, capacitance opposes the change and delays the voltage increase across the circuit. If applied voltage is decreased, capacitance tends to maintain the higher original voltage across the circuit, thus delaying the decrease.

Capacitance is also defined as that property of a circuit that enables energy to be stored in an electric field. Natural capacitance exists in many electric circuits. However, in this chapter, we are concerned only with the capacitance that is designed into the circuit by means of devices called *capacitors*.

✔ **Key Point:** The most noticeable effect of capacitance in a circuit is that voltage can neither increase nor decrease rapidly in a capacitive circuit, as it can in a circuit that does not include capacitance.

THE CAPACITOR

A *capacitor*, or condenser, is a manufactured electrical device that consists of two conducting plates of metal separated by an insulating material called a *dielectric* (see figure 9.73). (Note: the prefix *di-* means "through" or "across.")

The schematic symbol for a capacitor is shown in figure 9.74.

When a capacitor is connected to a voltage source, there is a short current pulse. A capacitor stores this electric charge in the dielectric (it can be charged and discharged, as we shall see later). To form a capacitor of any appreciable value, however,

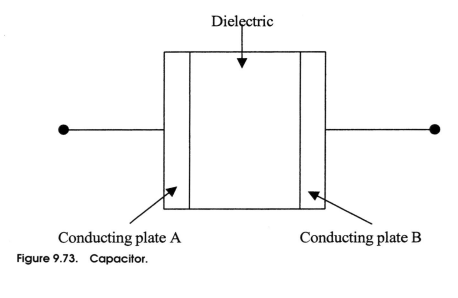

Figure 9.73. Capacitor.

Figure 9.74. (A) Schematic for a fixed capacitor. (B) Variable capacitor.

the area of the metal pieces must be quite large and the thickness of the dielectric must be quite small.

✔ **Key Point:** A capacitor is essentially a device that stores electrical energy.

The capacitor is used in a number of ways in electrical circuits. It may block d.c. portions of a circuit, since it is effectively a barrier to direct current (but not to a.c. current). It may be part of a tuned circuit—one such application is in the tuning of a radio to a particular station. It may be used to filter a.c. out of a d.c. circuit. Most of these are advanced applications that are beyond the scope of this book; however, a basic understanding of capacitance is necessary to the fundamentals of a.c. theory.

The two plates of the capacitor shown in figure 9.75 are electrically neutral since there are as many protons (positive charge) as electrons (negative charge) on each plate. Thus the capacitor has *no charge*.

Now a battery is connected across the plates (see figure 9.76 [A]). When the switch is closed (see figure 9.76 [B]), the negative charge on plate A is attracted to the positive terminal of the battery. This movement of charges will continue until the difference in charge between plates A and B is equal to the electromotive force (voltage) of the battery.

The capacitor is now *charged*. Because almost none of the charge can cross the space between plates, the capacitor will remain in this condition even if the battery is removed (see figure 9.77 [A]). However, if a conductor is placed across the plates (see figure 9.77 [B]), the electrons find a path back to plate A and the charges on each plate are again neutralized. The capacitor is now *discharged*.

✔ **Important Point:** In a capacitor, electrons cannot flow through the dielectric, because it is an insulator. Because it takes a definite quantity of electrons to

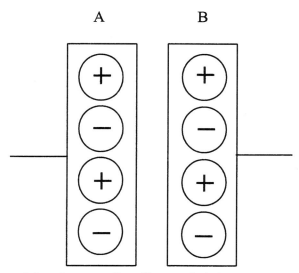

Figure 9.75. Two plates of a capacitor with a neutral charge.

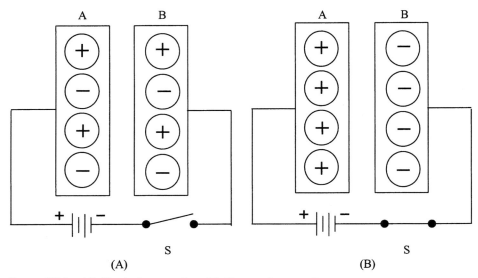

Figure 9.76.   (A) Neutral capacitor. (B) Charged capacitor.

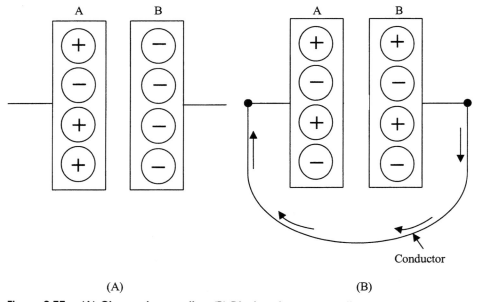

(A)   (B)

Figure 9.77.   (A) Charged capacitor. (B) Discharging a capacitor.

charge ("fill up") a capacitor, it is said to have *capacity*. This characteristic is referred to as *capacitance*.

DIELECTRIC MATERIALS

Somewhat similar to the phenomenon of permeability in magnetic circuits, various materials differ in their ability to support electric flux (lines of force) or to serve as dielectric material for capacitors. Materials are rated in their ability to support electric flux in terms of a number called a *dielectric constant*. Other factors being equal, the higher the value of the dielectric constant, the better is the dielectric material. Dry air is the standard (the reference) by which other materials are rated. Dielectric constants for some common materials are given in table 9.6.

✔ **Note:** From table 9.6 it is obvious that pure water is the best dielectric. Keep in mind that the key word is "pure." Water capacitors are used today in some high-energy applications, in which differences in potential are measured in thousands of volts.

UNIT OF CAPACITANCE

Capacitance is equal to the amount of charge that can be stored in a capacitor divided by the voltage applied across the plates:

$$C = \frac{Q}{E} \qquad (9.36)$$

where

C = capacitance, F (farads)
Q = amount of charge, C (coulombs)
E = voltage, V

Table 9.6   Dielectric constants

Material	Constant
Vacuum	1.0000
Air	1.0006
Paraffin paper	3.5
Glass	5–10
Quartz	3.8
Mica	3–6
Rubber	2.5–35
Wood	2.5–8
Porcelain	5.1–5.9
Glycerine (15°C)	56
Petroleum	2
Pure water	81

Example 9.28
Problem
What is the capacitance of two metal plates separated by 1 centimeter of air, if 0.002 coulomb of charge is stored when a potential of 300 volts is applied to the capacitor?

Solution
Given:

Q = 0.001 coulomb
F = 200 volts

$$C = \frac{Q}{E}$$

Converting to power of ten:

$$C = \frac{10 \times 10^{-4}}{2 \times 10^2}$$

$$C = 5 \times 10^{-6}$$
$$C = 0.000005 \text{ farads}$$

✔ Note: Although the capacitance value obtained in example 9.28 appears small, many electronic circuits require capacitors of much smaller value. Consequently the farad is a cumbersome unit, far too large for most applications. The *microfarad*, which is one-millionth of a farad (1×10^{-6} farad), is a more convenient unit. The symbol used to designated microfarad is μf.

Equation 9.36 can be rewritten as follows:

$$Q = CE \tag{9.37}$$

$$E = \frac{Q}{C} \tag{9.38}$$

✔ Important Point: From equation 9.37, do not deduce the mistaken idea that capacitance is dependent upon charge and voltage. Capacitance is determined entirely by physical factors. The symbol used to designate a capacitor is (C). The unit of capacitance is the farad (F). The farad is that capacitance that will store 1 coulomb of charge in the dielectric when the voltage applied across the capacitor terminals is 1 volt.

FACTORS AFFECTING THE VALUE OF CAPACITANCE

The capacitance of a capacitor depends on three main factors: plate surface area; distance between plates; and dielectric constant of the insulating material.

• **Plate surface area**: Capacitance varies directly with place surface area. We can double the capacitance value by doubling the capacitor's plate surface area. Figure 9.78 shows a capacitor with a small surface area and another one with a large surface area.

Adding more capacitor plates can increase the plate surface area. Figure 9.79 shows alternate plates connecting to opposite capacitor terminals.

• **Distance between plates**: Capacitance varies inversely with the distance between plate surfaces. The capacitance increases when the plates are closer together. Figure 9.80 shows capacitors with the same plate surface area, but with different spacing.

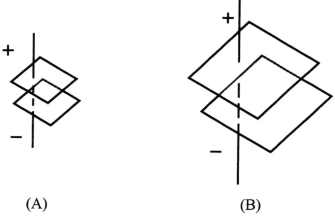

(A)   (B)

Figure 9.78.   (A) Small plates, small capacitance. (B) Larger plates, higher capacitance.

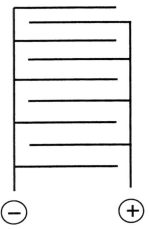

Figure 9.79.   Several sets of plates connected to produce a capacitor with more surface area.

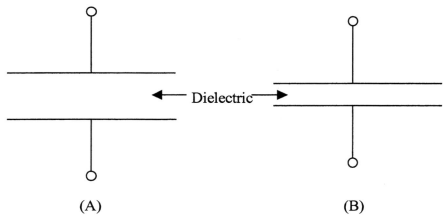

Figure 9.80. (A) Wide plate spacing, small capacitance. (B) Narrow plate spacing, larger capacitance.

- **Dielectric constant of the insulating material**: An insulating material with a higher dielectric constant produces a higher capacitance rating. Figure 9.81 shows two capacitors. Both have the same plate surface area and spacing. Air is the dielectric in the first capacitor and mica is the dielectric in the second one. Mica's dielectric constant is 5.4 times greater than air's dielectric constant. The mica capacitor has 5.4 times more capacitance than the air-dielectric capacitor.

VOLTAGE RATING OF CAPACITORS

There is a limit to the voltage that may be applied across any capacitor. If too large a voltage is applied, it will overcome the resistance of the dielectric and a current will be forced through it from one plate to the other, sometimes burning a hole in the dielectric. In this event, a short circuit exists and the capacitor must be discarded. The maximum voltage that may be applied to a capacitor is known as the *working voltage* and must never be exceeded.

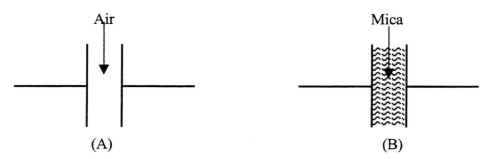

Figure 9.81. (A) Low capacitance. (B) Higher capacitance.

The working voltage of a capacitor depends on (1) the type of material used as the dielectric, and (2) the thickness of the dielectric. As a margin of safety, the capacitor should be selected so that its working voltage is at least 50% greater than the highest voltage to be applied to it. For example, if a capacitor is expected to have a maximum of 200 volts applied to it, its working voltage should be at least 300 volts.

CAPACITORS IN SERIES AND PARALLEL

Like resistors or inductors, capacitors may be connected in series, in parallel, or in a series-parallel combination. Unlike resistors or inductors, however, total capacitance in series, in parallel, or in a series-parallel combination is found in a different manner. Simply put, the rules are not the same for the calculation of total capacitance. This difference is explained as follows:

Parallel capacitance is calculated like series resistance, and series capacitance is calculated like parallel resistance. For example:

When capacitors are connected in *series* (see figure 9.82), the total capacitance C_T is

$$\text{Series:} \quad \frac{1}{C_T} = \frac{1}{C_1} + \frac{1}{C_2} + \frac{1}{C_3} + \ldots + \frac{1}{C_n} \tag{9.39}$$

Example 9.29
Problem
Find the total capacitance of a 3 μF, a 5 μF, and a 15 μF capacitor in series.

Solution
Write equation 9.39 for three capacitors in series.

$$\frac{1}{C_T} = \frac{1}{C_1} + \frac{1}{C_2} + \frac{1}{C_3}$$

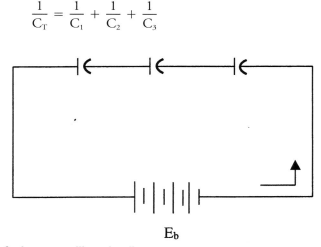

E_b

Figure 9.82.   Series capacitive circuit.

$$= \frac{1}{3} + \frac{1}{5} + \frac{1}{15} = \frac{9}{15} = \frac{3}{5} = \frac{5}{3} = 1.7 \ \mu F$$

When capacitors are connected in parallel (see figure 9.83), the total capacitance C_T is the sum of the individual capacitances.

$$\text{Parallel:} \quad C_T = C_1 + C_2 + C_3 + \ldots + C_n \qquad (9.40)$$

Example 9.30
Problem
Determine the total capacitance in a parallel capacitive circuit.

Given:

$C_1 = 2 \ \mu F$
$C_2 = 3 \ \mu F$
$C_3 = 0.25 \ \mu F$

Solution
Write equation 9.40 for three capacitors in parallel:

$$\begin{aligned} C_T &= C_1 + C_2 + C_3 \\ &= 2 + 3 + 0.25 \\ &= 5.25 \ \mu F \end{aligned}$$

If capacitors are connected in a combination of *series and parallel* (see figure 9.84), the total capacitance is found by applying equations 9.5 and 9.6 to the individual branches.

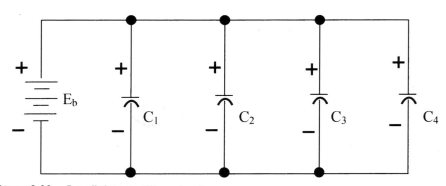

Figure 9.83.   Parallel capacitive circuit.

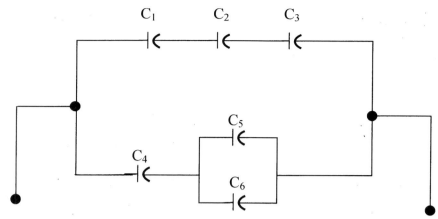

Figure 9.84. Series-parallel capacitance configuration.

TYPES OF CAPACITORS

Capacitors used for commercial applications are divided into two major groups—fixed and variable—and are named according to their dielectric. Most common are air, mica, paper, and ceramic capacitors, plus the electrolytic type. These types are compared in table 9.7.

The fixed capacitor has a set value of capacitances that is determined by its construction. The construction of the variable capacitor allows a range of capacitances. Within this range, the desired value of capacitance is obtained by some mechanical means, such as by turning a shaft (as in turning a radio tuner knob, for example) or adjusting a screw to adjust the distance between the plates.

The electrolytic capacitor consists of two metal plates separated by an electrolyte. The electrolyte, either paste or liquid, is in contact with the negative terminal, and this combination forms the negative electrode. The dielectric is a very thin film of oxide deposited on the positive electrode, which is aluminum sheet. Electrolytic capacitors are polarity sensitive (i.e., they must be connected in a circuit according to their polarity markings)—used where a large amount of capacitance is required.

Table 9.7 Comparison of capacitor types

Dielectric	Construction	Capacitance range
Air	Meshed plates	10–400 pF
Mica	Stacked plates	10–5000 pF
Paper	Rolled foil	0.001–1 µF
Ceramic	Tubular	0.5–1600 pF
	Disk	0.002–0.1 µF
Electrolytic	Aluminum	5–1000 µF
	Tantalum	0.01–300 µF

Inductive and Capacitive Reactance

We have learned that the inductance of a circuit acts to oppose any change of current flow in that circuit and that capacitance acts to oppose any change of voltage. In d.c. circuits these reactions are not important, because they are momentary and occur only when a circuit is first closed or opened. In a.c. circuits these effects become very important because the direction of current flow is reversed many times each second, and the opposition presented by inductance and capacitance is, for practical purposes, constant.

In purely resistive circuits, either d.c. or a.c., the term for opposition to current flow is *resistance*. When the effects of capacitance or inductance are present, as they often are in a.c. circuits, the opposition to current flow is called *reactance*. The total opposition to current flow in circuits that have both resistance and reactance is called *impedance*.

INDUCTIVE REACTANCE

In order to gain understanding of the reactance of a typical coil, we need to review exactly what occurs when an a.c. voltage is impressed across the coil.

1. The a.c. voltage produces an alternating current.
2. When a current flows in a wire, lines of force are produced around the wire.
3. Large currents produce many lines of force; small currents produce only a few lines of force.
4. As the current changes, the number of lines of force will change. The field of force will seem to expand and contract as the current increases and decreases as shown in figure 9.85.

Expanding lines of force

Contracting lines of force

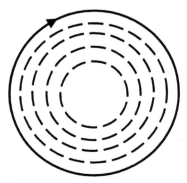
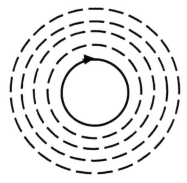

Figure 9.85.   Alternating current producing a moving (expanding and collapsing) field. In a coil, this moving field cuts the wires of the coil.

5. As the field expands and contracts, the lines of force must cut across the wires that form the turns of the coil.
6. These cuttings induce an emf in the coil.
7. This emf acts in the direction so as to oppose the original voltage and is called a *counter*, or *back*, *emf*.
8. The effect of this counter emf is to reduce the original voltage impressed on the coil. The net effect will be to reduce the current below that which would flow if there were no cuttings or counter emf.
9. In this sense, the counter emf is acting as a resistance in reducing the current.
10. Although it would be more convenient to consider the current-reducing effect of a counter emf as a number of ohms of effective resistance, we don't do this. Instead, since a counter emf is not actually a resistance but merely *acts* as a resistance, we use the term *reactance* to describe this effect.

▶ **Important Point:** The reactance of a coil is the number of ohms of resistance that the coil *seems* to offer as a result of a counter emf induced in it. Its symbol is X, to differentiate it from the d.c. resistance *R*.

Inductive reactance of a coil depends primarily on (1) the coil's inductance and (2) the frequency of the current flowing through the coil. The value of the reactance of a coil is therefore proportional to its inductance and the frequency of the a.c. circuit in which it is used.

The formula for inductive reactance is

$$X_L = 2\pi fL \tag{9.41}$$

Since $2\pi = 2(3.14) = 6.28$, equation 9.41 becomes

$$X_L = 6.28fL$$

where

X_L = inductive reactance, Ω
f = frequency, Hz
L = inductance, H

If any two quantities are known in equation 9.41, the third can be found.

$$L = \frac{X_L}{6.28\ f} \tag{9.42}$$

$$f = \frac{X_L}{6.28\ L} \tag{9.43}$$

Example 9.31
Problem
The frequency of a circuit is 60 Hz and the inductance is 20 mh. What is X_L?

Solution

$$X_L = 2\pi fL$$
$$= 6.28 \times 60 \times 0.02$$
$$= 7.5 \ \Omega$$

Example 9.32
Problem
A 30-mh coil is in a circuit operating at a frequency of 1,400 kHz. Find its inductive reactance.

Solution
Given:

L = 30 mh
f = 1,400 kHz

Find X_L.

Step 1: Change units of measurement.

$$30 \text{ mH} = 30 \times 10^{-3} \text{ h}$$
$$1,400 \text{ kHz} = 1,400 \times 10^3$$

Step 2: Find the inductive reactance.

$$X_L = 6.28fL$$
$$X_L = 6.28 \times 1,400 \times 10^3 \times 30 \times 10^{-3}$$
$$X_L = 263,760 \ \Omega$$

Example 9.33
Problem
Given:

L = 400 μh
f = 1,500 Hz

Find X_L.

Solution

$$X_L = 2\pi fL$$
$$= 6.28 \times 1,500 \times 0.0004$$
$$= 3.78 \ \Omega$$

✔ **Key Point:** If frequency or inductance varies, inductive reactance must also vary. A coil's inductance does not vary appreciably after the coil is manufactured, unless it is designed as a variable inductor. Thus, frequency is generally the only variable factor affecting the inductive reactance of a coil. The coil's inductive reactance will vary directly with the applied frequency.

CAPACITIVE REACTANCE

Previously, we learned that as a capacitor is charged, electrons are drawn from one plate and deposited on the other. As more and more electrons accumulate on the second plate, they begin to act as an opposing voltage, which attempts to stop the flow of electrons just as a resistor would do. This opposing effect is called the *reactance* of the capacitor and is measured in ohms. The basic symbol for reactance is X, and the subscript defines the type of reactance. In the symbol for inductive reactance, X_L, the subscript L refers to inductance. Following the same pattern, the symbol for capacitive reactance is X_C.

✔ **Key Point:** *Capacitive reactance*, X_C, is the opposition to the flow of a.c. current due to the capacitance in the circuit.

Factors affecting capacitive reactance, X_C, are:

- the size of the capacitor
- frequency

The larger the capacitor, the greater the number of electrons that may be accumulated on its plates. However, because the plate area is large, the electrons do not accumulate in one spot but spread out over the entire area of the plate and do not impede the flow of new electrons onto the plate. Therefore, a large capacitor offers a small reactance. If the capacitance were small, as in a capacitor with a small plate area, the electrons could not spread out and would attempt to stop the flow of electrons coming to the plate. Therefore, a small capacitor offers a large reactance. The reactance is therefore *inversely* proportional to the capacitance.

If an a.c. voltage is impressed across the capacitor, electrons are accumulated first on one plate and then on the other. If the frequency of the changes in polarity is low, the time available to accumulate electrons will be large. This means that a large number of electrons will be able to accumulate, which will result in a large opposing effect, or a large reactance. If the frequency is high, the time available to accumulate electrons will be small. This means that there will be only a few electrons on the plates, which will result in a small opposing effect, or a small reactance. The reactance is, therefore, *inversely* proportional to the frequency.

The formula for capacitive reactance is

$$X_C = \frac{1}{2\pi fC} \qquad\qquad (9.44)$$

with C measured in farads.

Example 9.34
Problem
What is the capacitive reactance of a circuit operating at a frequency of 60 Hz, if the total capacitance is 130 μf?

Solution

$$X_C = \frac{1}{2\pi fC}$$

$$= \frac{1}{6.28 \times 60 \times 0.00013}$$

$$= 20.4 \ \Omega$$

PHASE RELATIONSHIP R, L, AND C CIRCUITS

Unlike a purely resistive circuit (where current rises and falls with the voltage; that is, it neither leads nor lags and current and voltage are in phase), current and voltage are not in phase in inductive and capacitive circuits. This is the case, of course, because occurrences are not quite instantaneous in circuits that have either inductive or capacitive components.

In the case of an inductor, voltage is first applied to the circuit, then the magnetic field begins to expand, and self-induction causes a counter current to flow in the circuit, opposing the original circuit current. *Voltage leads current* by 90 degrees (see figure 9.86).

When a circuit includes a capacitor, a charge current begins to flow and then a difference in potential appears between the plates of the capacitor. *Current leads voltage* by 90 degrees (see figure 9.87).

✔ **Key Point:** In an inductive circuit, *voltage leads current* by 90 degrees; and in a capacitive circuit, *current leads voltage* by 90 degrees.

IMPEDANCE

Impedance is the total opposition to the flow of alternating current in a circuit that contains resistance and reactance. In the case of pure inductance, inductive reactance

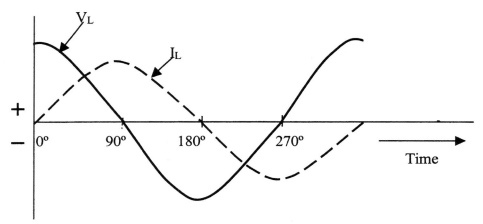

Figure 9.86.   Inductive circuit: voltage leads current by 90°.

(X_L) is the total opposition to the flow of current through it. In the case of pure resistance, R represents the total opposition. The combined opposition of R and X_L in series or in parallel to current flow is called impedance. The symbol for impedance is Z.

The impedance of resistance in series with inductance is

$$Z = \sqrt{R^2 + X_L^2} \tag{9.45}$$

where

Z = Impedance, Ω
R = Resistance, Ω
X_L = Inductive reactance, Ω

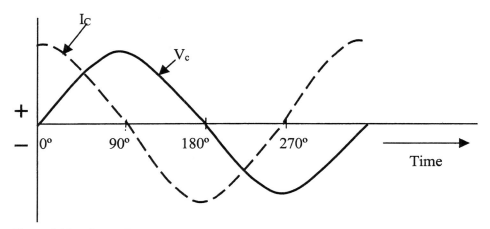

Figure 9.87.   Capacitive circuit: current leads voltage by 90°.

The impedance of resistance in series with capacitance is:

$$Z = \sqrt{R^2 + X_C^2}$$ (9.46)

where

Z = Impedance, Ω
R = Resistance, Ω
X_C = Inductive capacitance, Ω

When the impedance of a circuit includes R, X_L, and X_C, both resistance and net reactance must be taken into account. The equation for impedance, including both X_L and X_C, is:

$$Z = \sqrt{R^2 + (X_L - X_C)^2}$$ (9.47)

POWER IN REACTIVE CIRCUITS

The power in a d.c. circuit is equal to the product of volts and amps, but in an a.c. circuit this is true only when the load is resistive and has no reactance.

In a circuit possessing inductance only, the true power is zero. The current lags the applied voltage by 90 degrees. The true power in a capacitive circuit is also zero. The *true power* is the average power actually consumed by the circuit, the average being taken over one complete cycle of alternating current. The *apparent power* is the product of the rms volts and rms amps.

The ratio of true power to apparent power in an a.c. circuit is called the *power factor*. It may be expressed as a percentage or as a decimal.

A.C. Circuit Theory

To this point we have explained how a combination of inductance and resistance and then capacitance and resistance behave in an a.c. circuit. We saw how the RL and RC combination affects the current, voltages, power, and power factor of a circuit. We considered these fundamental properties as isolated phenomena. The following phase relationships were seen to be true:

1. The voltage drop across a resistor is *in phase* with the current through it.
2. The voltage drop across an inductor *leads* the current through it by 90°.
3. The voltage drop across a capacitor *lags* the current through it by 90°.
4. The voltage drops across inductors and capacitors are *180 degrees out of phase*.

Solving a.c. problems is complicated by the fact that current varies with time as the a.c. output of an alternator goes through a complete cycle. This is the case because the various voltage drops in the circuit vary in-phase—they are not at their maximum or minimum values at the same time.

A.C. circuits frequently include all three circuit elements: resistance, inductance, and capacitance.

SERIES RLC CIRCUIT

Figure 9.88 shows both the sine waveforms and the vectors for purely resistive, inductive, and capacitive circuits. Only the vectors show the direction, because the magnitudes are dependent on the values chosen for a given circuit. [**Note:** We are only interested in the *effective* (root-mean square, rms) values.] If the individual resistances and reactances are known, Ohm's law may be applied to find the voltage drops. For example, we know that $E_R = I \times R$ and $E_C = I \times X_L$. Then, according to Ohm's law, $E_L = I \times X_L$.

In an a.c. circuit, current varies with time; accordingly, the voltage drops across the various elements also vary with time. However, the same variation is not always present in each *at the same time* (except in purely resistive circuits) because current and voltage are not in phase.

✔ **Important Point:** In a resistive circuit, the phase difference between voltage and current is zero.

We are concerned, in practical terms, mostly with effective values of current and voltage. However, to understand basic a.c. theory, we need to know what occurs from instant to instant.

In figure 9.89, note first that current is the common reference for all three element voltages, because there is only one current in a series circuit, and it is common to all elements. The dashed line in figure 9.89 (A) represents the common series current. The voltage vector for each element, showing its individual relation to the common current, is drawn above each respective element. The total source voltage E is the vector sum of the individual voltages of IR, IX_L, and IX_C.

The three element voltages are arranged for summation in figure 9.89 (B). Because IX_L and IX_C are each 90° away from I, they are therefore 180° from each other. Vectors in direct opposition (180° out of phase) may be subtracted directly. The total reactive voltage E_X is the difference of IX_L and IX_C. Or, $E_X = IX_L - IX_C = 45 - 15 = 30$ volts.

✔ **Important Point:** The voltage across a single reactive element in a series circuit can have a greater effective value than that of the applied voltage.

The final relationship of line voltage and current, as seen from the source, is shown in figure 9.89 (C). Had X_C been larger than X_L, the voltage would lag rather than lead. When X_C and X_L are of equal values, line voltage and current will be in phase.

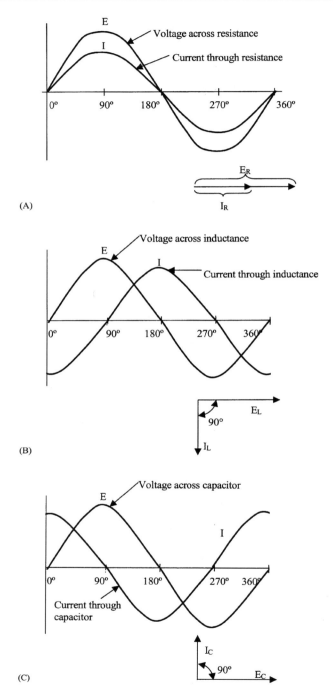

Figure 9.88.   Sine waveforms and vector representation of R, L, and C circuits. (A) Pure resistive circuit (voltage and current are in phase). (B) Pure inductive circuit (voltage leads current by 90°). (C) Pure capacitive circuit (voltage lags current by 90°).

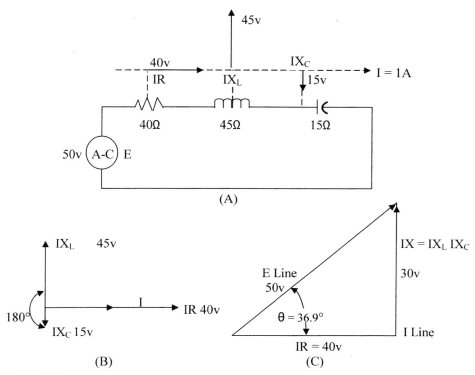

Figure 9.89.   Resistance, inductance, and capacitance connected in a series.

✔ **Important Point:** One of the most important characteristics of an RLC circuit is that it can be made to respond most effectively to a single given frequency. When operated in this condition, the circuit is said to be in *resonance* with or *resonant* to the operating frequency. A circuit is at resonance when the inductive reactance X_L is equal to the capacitive reactance X_C. At resonance Z equals the resistance R.

In summary, the series RLC circuit illustrates three important points:

• The current in a series RLC circuit either leads or lags the applied voltage, depending on whether X_C is greater or less than X_L.
• A capacitive voltage drop in a series circuit always subtracts directly from an inductive voltage drop.
• The voltage across a single reactive element in a series circuit can have a greater effective value than that of the applied voltage.

PARALLEL RLC CIRCUITS

The *true power* of a circuit is $P = EI \cos \theta$; for any given amount of power to be transmitted, the current, I, varies inversely with the power factor, $\cos \theta$. Thus, the addition of capacitance in parallel with inductance will, under the proper conditions, improve the power factor (make nearer to unity power factor) of the circuit and make possible the transmission of electric power with reduced line loss and improved voltage regulation.

Figure 9.90 (A) shows a three-branch parallel a.c. circuit with a resistance in one branch, inductance in the second branch, and capacitance in the third branch. The voltage is the same across each parallel branch, so $V_T = V_R = V_L = V_C$. The applied voltage V_T is used as the reference line to measure phase angle θ. The total current I_T is the vector sum of I_R, I_L, and I_C. The current in the resistance I_R is in phase with the applied voltage V_T (see figure 9.90 [B]). The current in the capacitor I_C leads the voltage V_T by 90°. I_L and I_C are exactly 180° out of phase and thus acting in opposite directions)see figure 9.90 [B]). When $I_C > I_C$, I_T lags V_T (see figure 9.90 [C]), the parallel RLC circuit is considered inductive.

POWER IN A.C. CIRCUITS

In circuits that have only resistance but no reactance, the amount of power absorbed in the circuit is easily calculated by $P = I^2 R$. However, in dealing with circuits that include inductance and capacitance (or both), which is often the case in a.c. electricity, the calculation of power is a more complicated process.

Earlier, we explained that power is a measure of the rate at which work is done. The "work" of a resistor is to limit current flow to the correct, safe level. In accomplishing this, the resistor dissipates heat, and we say that power is consumed or absorbed by the resistor.

Inductors and capacitors also oppose current flow, but they do so by producing current that opposes the line current. In either inductive or capacitive circuits, instantaneous values of power may be very large, but the power actually absorbed is essentially zero, because only resistance dissipates heat (absorbs power). Both inductance and capacitance return the power to the source.

Any component that has resistance, such as a resistor or the wiring of an inductor,

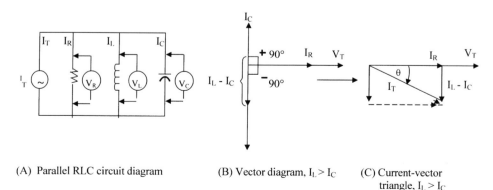

(A) Parallel RLC circuit diagram   (B) Vector diagram, $I_L > I_C$ (C) Current-vector
                                                                                triangle, $I_L > I_C$

Figure 9.90.   R, X$_L$, and X$_C$ in parallel: parallel RLC circuit diagram; vector diagram, I$_L$ > I$_C$; current-vector triangle, I$_L$ > I$_C$.

consumes power. Such power is not returned to the source, because it is dissipated as heat. Previously, we stated that power consumed in the circuit is called *true power*, or *average power*. The two terms are interchangeable, but we use the term *average power*, because the overall value is more meaningful than the instantaneous values of power appearing in the circuit during a complete cycle.

✔ **Key Point:** In terms of the dissipation of power as heat in a circuit, *apparent power* includes both power that is returned to the source and power that is dissipated as heat. *Average power* is power that is dissipated as heat.

Not all apparent power is consumed by the circuit; however, because the alternator does deliver the power, it must be considered in the design. The average power consumption may be small, but instantaneous values of voltage and current are often very large. Apparent power is an important design consideration, especially in assessing the amount of insulation necessary.

In an a.c. circuit that includes both reactance and resistance, some power is consumed by the load and some is returned to the source. How much of each depends on the phase angle, since current normally leads or lags voltage by some angle.

✔ **Note:** Recall that in a purely reactive circuit, current and voltage are 90° out of phase.

In an RLC circuit, the ratio of R/Z is the cosine of the phase angle θ. Therefore, it is easy to calculate average power in an RLC circuit:

$$P = EI \cos \theta \tag{9.48}$$

where

E = effective value of the voltage across the circuit
I = effective value of current in the circuit
θ = phase angle between voltage and current
P = average power absorbed by the circuit

✔ **Note:** Recall that the equation for average power in a purely resistive circuit is P = EI. In a resistive circuit, P = EI, because the cos θ is 1 and need not be considered. In most cases, the phase angle will be neither 90° nor zero, but somewhere between those extremes.

Example 9.35
Problem
An RLC circuit has a source voltage of 500 volts, line current is 2 amps, and current leads voltage by 60 degrees. What is the average power?

Solution

$$\text{Average power} = 500 \text{ v} \times 2 \text{ a} \times 0.5 \text{ (Note: Cos of } 60° = 0.5)$$
$$= 500 \text{ watts}$$

Example 9.36
Problem
An RLC circuit has a source voltage of 300 volts, line current is 2 amps, and current lags voltage by 31.8°. What is the average power?

Solution

$$\text{Average power} = 300 \text{ v} \times 2 \text{ a} \times 0.8499$$
$$= 509.9 \text{ watts}$$

Example 9.37
Problem
Given:

E = 100 volts
I = 4 amps
θ = 58.4

What is the average power?

Solution

$$\text{Average power} = 100 \text{ v} \times 4 \text{ a} \times 0.5240$$
$$= 209.6 \text{ watts}$$

Electrical Applications

To this point in the chapter, we have concentrated (in brief fashion, of course) on the fundamentals of electricity and electric circuits. This was our goal. Moreover, along with satisfying our goal, we also understand that having a basic knowledge of electrical theory is a great accomplishment. However, a knowledge of basic theory (of any type) that is not put to practical use is analogous to understanding the operation of an internal combustion engine without ever having the opportunity to work on one.

ELECTRICAL POWER GENERATION

Generators can be designed to supply small amounts of power or they can be designed to supply many thousands of kilowatts of power. Also, generators may be designed to supply either direct current or alternating current.

D.C. Generators

A *d.c. generator* is a rotating machine that converts mechanical energy into electrical energy. This conversion is accomplished by rotating an armature, which carries conductors, in a magnetic field, thus inducing an emf in the conductors. As stated previously, in order for an emf to be induced in the conductors, a relative motion must always exist between the conductors and the magnetic field in such a manner that the conductors cut through the field. In most d.c. generators, the armature is the rotating member and the field is the stationary member. A mechanical force is applied to the shaft of the rotating member to cause the relative motion. Thus, when mechanical energy is put into the machine in the form of a mechanical force or twist on the shaft, causing the shaft to turn at a certain speed, electrical energy in the form of voltage and current is delivered to the external load circuit.

✔ **Important Point:** Mechanical power must be applied to the shaft constantly as long as the generator is supplying electrical energy to the external load circuit.

To gain a basic understanding of the operation of a d.c. generator, consider the following explanation.

A simple d.c. generator consists of an armature coil with a single turn of wire (see figure 9.91 [A] and [B]). (Note: The armature coils used in large d.c. machines are usually wound in their final shape before being put on the armature. The sides of the preformed coil are placed in the slots of the laminated armature core.) This armature coil cuts across the magnetic field to produce voltage. If a complete path is present, current will move through the circuit in the direction shown by the arrows (see figure 9.91 [A]). In this position of the coil, commutator segment 1 is in contact with brush 1, while commutator segment 2 is in contact with brush 2. As the armature rotates a half turn in a clockwise direction, the contacts between the commutator segments and the brushes are reversed (see figure 9.91 [B]). At this moment, segment 1 is in contact with brush 2 and segment 2 is in contact with brush 1. Because of this commutator

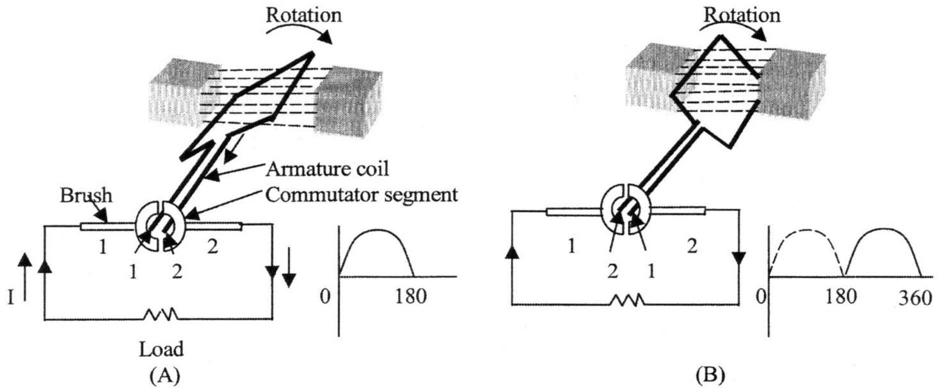

Figure 9.91. Basic operation of a d.c. generator.

action, that side of the armature coil that is in contact with either of the brushes is always cutting across the magnetic field in the same direction. Thus, brushes 1 and 2 have constant polarity, and a *pulsating d.c. current* is delivered to the external load circuit.

✔ **Note:** In d.c. generators, voltage induced in individual conductors is a.c. It is converted to d.c. (rectified) by the commutator, which rotates in contact with carbon brushes so that the current generated is in one direction, or direct current.

There are several different types of d.c. generators. They take their names from the type of field excitation used (i.e., they are classified according to the manner in which the field windings are connected to the armature circuit). For example, when the generator's field is excited (or supplied) from a separate d.c. source (such as a battery) other than its own armature, it is called a separately excited d.c. generator (see figure 9.92).

A *shunt generator* (self-excited) has its field windings connected in series with a rheostat, across the armature in shunt with the load, as shown in figure 9.93. The shunt generator is widely used in industry.

A series generator (self-excited) has its field windings connected in series with the armature and load, as shown in figure 9.94. Series generators are seldom used.

Compound generators (self-excited) contain both series and shunt field windings, as shown in figure 9.95. Compound generators are widely used in industry.

✔ **Note:** As central generating stations increased in size along with number and power distribution distances, d.c. generating systems, because of the high power losses in long d.c. transmission lines, were replaced by a.c. generating systems to reduce power transmission costs.

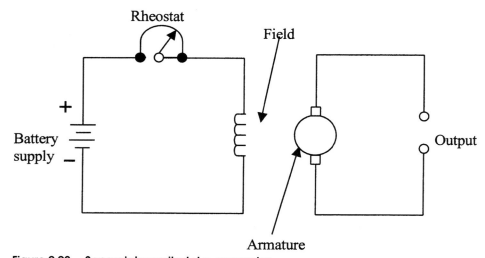

Figure 9.92. Separately excited d.c. generator.

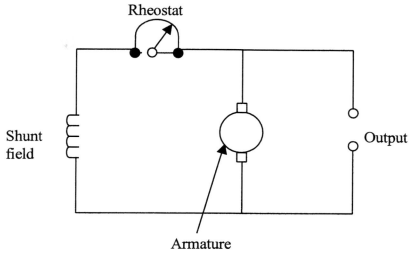

Figure 9.93.   D.C. shunt generator.

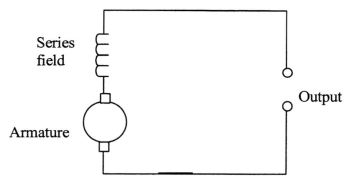

Figure 9.94.   D.C. series generator.

A.C. Generators

Most electric power utilized today is generated by *alternating-current generators* (also called *alternators*). They are made in many different sizes, depending on their intended use. Regardless of size, however, all generators operate on the same basic principle—a magnetic field cutting through conductors, or conductors passing through a magnetic field. They are (1) a group of conductors in which the output voltage is generated, and (2) a second group of conductors through which direct current is passed to obtain an electromagnetic field of fixed polarity. The conductors in which the electromagnetic field originates are always referred to as the field windings.

In addition to the armature and field, there must also be motion between the two. To provide this, a.c. generators are built in two major assemblies, the *stator* and the *rotor*. The rotor rotates inside the stator.

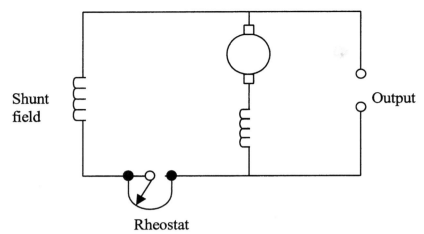

Figure 9.95.   D.C. compound generator.

The revolving-field a.c. generator (see figure 9.96) is the most widely used type. In this type of generator, direct current from a separate source is passed through windings on the rotor by means of sliprings and brushes. (Note: Sliprings and brushes are adequate for the d.c. field supply because the power level in the field is much smaller than in the armature circuit.) This maintains a rotating electromagnetic field of fixed polarity. The rotating magnetic field, following the rotor, extends outward and cuts through the armature windings imbedded in the surrounding stator. As the rotor turns, a.c. voltages are induced in the windings, since magnetic fields of first one polarity and then the other cut through them. Because the output power is taken from stationary windings, the output may be connected through fixed output terminals T1

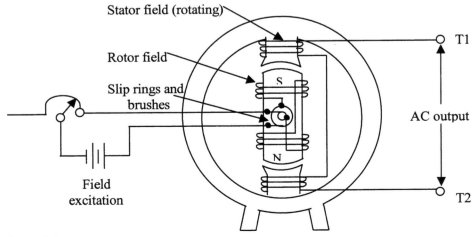

Figure 9.96.   Essential parts of a rotating field a.c. generator.

and T2 in figure 9.96. This is advantageous, in that there are no sliding contacts and the whole output circuit is continuously insulated.

✔ **Important Point:** In a.c. generators, frequency and electromagnetic wave cycles per second depend on how fast the rotor turns and the number of electromagnetic field poles. Voltage generated depends on the rotor speed, number of coils in the armature and strength of the magnetic field.

MOTORS

At least 60% of the electrical power fed to a typical industrial plant is consumed by electric motors. One thing is certain: there is an almost endless variety of tasks that electric motors perform in industry.

An *electric motor* is a machine used to change electrical energy to mechanical energy to do the work. (Note: Recall that a generator does just the opposite; that is, a generator changes mechanical energy to electrical energy.)

Previously, we pointed out that when a current passes through a wire, a magnetic field is produced around the wire. If this magnetic field passes through a stationary magnetic field, the fields either repel or attract, depending on their relative polarity. If both are positive or both are negative, they repel. If they are opposite polarity, they attract.

Applying this basic information to motor design, an electromagnetic coil, the armature, rotates on a shaft. The armature and shaft assembly are called the rotor. The rotor is assembled between the poles of a permanent magnet, and each end of the rotor coil (armature) is connected to a commutator also mounted on the shaft. A commutator is composed of copper segments insulated from the shaft and from each other by an insulating material. As like poles of the electromagnet in the rotating armature pass the stationary permanent magnet poles, they are repelled. As the opposite poles near each other, they attract, continuing the motion.

D.C. Motors

The construction of a d.c. motor is essentially the same as that of a d.c. generator. However, it is important to remember that the d.c. generator converts mechanical energy into electrical energy back into mechanical energy. A d.c. generator may be made to function as a motor by applying a suitable source of d.c. voltage across the normal output electrical terminals.

There are various types of d.c. motors, depending on the way the field coils are connected. Each has characteristics that are advantageous under given load conditions.

Shunt motors (see figure 9.97) have the field coils connected in parallel with the armature circuit. This type of motor, with constant potential applied, develops variable torque at an essentially constant speed, even under changing load conditions. Such loads are found in machine-shop equipment such as lathes, shapes, drills, milling machines, and so forth.

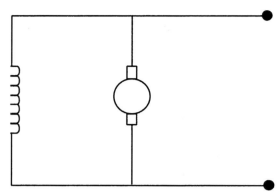

Figure 9.97.   D.C. shunt motor.

Series motors (see figure 9.98) have the field coils connected in series with the armature circuit. This type of motor, with constant potential applied, develops variable torque but its speed varies widely under changing load conditions. That is, the speed is low under heavy loads, but becomes excessively high under light loads. Series motors are commonly used to drive electric hoists, winches, cranes, and certain types of vehicles (e.g., electric trucks). Also, series motors are used extensively to start internal combustion engines.

Compound motors (see figure 9.99) have one set of field coils in parallel with the armature circuit, and another set of field coils in series with the armature circuit. This type of motor is a compromise between shunt and series motors. It develops an increased starting torque over that of the shunt motor, and has less variation in speed than the series motor.

The speed of a d.c. motor is variable. It is increased or decreased by a rheostat connected in series with the field or in parallel with the rotor. Interchanging either the rotor or field winding connections reverses direction.

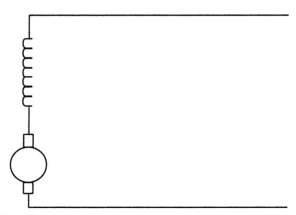

Figure 9.98.   D.C. series motor.

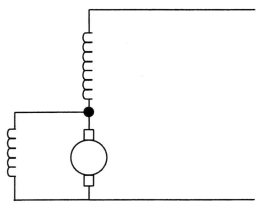

Figure 9.99.   D.C. compound motor.

A.C. Motors

A.C. voltage can be easily transformed from low voltages to high voltages or vice versa, and can be moved over a much greater distance without too much loss in efficiency. Most of the power-generating systems today, therefore, produce alternating current. Thus, it logically follows that a great majority of the electrical motors utilized today are designed to operate on alternating current. However, there are other advantages in the use of a.c. motors besides the wide availability of a.c. power. In general, a.c. motors are less expensive than d.c. motors. Most types of a.c. motors do not employ brushes and commutators. This eliminates many problems of maintenance and wear and also eliminates dangerous sparking.

A.C. motors are manufactured in many different sizes, shapes, and ratings, for use on an even greater number of jobs. They are designed for use with either polyphase or single-phase power systems.

This chapter cannot possibly cover all aspects of the subject of a.c. motors. Consequently, it deals mainly with the operating principles of the two most common a.c. motor types: the induction and the synchronous motor.

Induction Motors (Polyphase). The induction motor is the most commonly used type of a.c. motor because of its simple, rugged construction and good operating characteristics. It consists of two parts: the *stator* (stationary part) and the *rotor* (rotating part).

The most important type of polyphase induction motor is the three-phase motor.

✔ **Important Note:** A three-phase (3-θ) system is a combination of three single-phase (1-θ) systems. In a 3-θ balanced system, the power comes from an a.c. generator that produces three separate but equal voltages, each of which is out of phase with the other voltages by 120°. Although 1-θ circuits are widely used in electrical systems, most generation and distribution of a.c. current is 3-θ.

The driving torque of both d.c. and a.c. motors is derived from the reaction of current-carrying conductors in a magnetic field. In the d.c. motor, the magnetic field is stationary and the armature, with its current-carrying conductors, rotates. The current is supplied to the armature through a commutator and brushes.

In *induction motors*, the rotor currents are supplied by electromagnet induction. The stator windings, connected to the a.c. supply, contain two or more out-of-time-phase currents, which produce corresponding mmf. These mmf establish a rotating magnetic field across the air-gap. This magnetic field rotates continuously at constant speed regardless of the load on the motor. The stator winding corresponds to the armature winding of a d.c. motor or to the primary winding of a transformer. The rotor is not connected electrically to the power supply.

The induction motor derives its name from the fact that mutual induction (or transformer action) takes place between the stator and the rotor under operating conditions. The magnetic revolving field produced by the stator cuts across the rotor conductors, inducing a voltage in the conductors. This induced voltage causes rotor current to flow. Hence, motor torque is developed by the interaction of the rotor current and the magnetic revolving field.

Synchronous Motors. Like induction motors, *synchronous motors* have stator windings that produce a rotating magnetic field. But unlike the induction motor, the synchronous motor requires a separate source of d.c. from the field. It also requires special starting components. These include a salient-pole field with starting grid winding. The rotor of the conventional type synchronous motor is essentially the same as that of the salient-pole a.c. generator. The stator windings of induction and synchronous motors are essentially the same.

In operation, the synchronous motor rotor locks into step with the rotating magnetic field and rotates at the same speed. If the rotor is pulled out of step with the rotating stator field, no torque is developed and the motor stops. Because a synchronous motor develops torque only when running at synchronous speed, it is not self-starting and hence needs some device to bring the rotor to synchronous speed. For example, a synchronous motor may be started rotating with a d.c. motor on a common shaft. After the motor is brought to synchronous speed, a.c. current is applied to the stator windings. The d.c. starting motor now acts as a d.c. generator, which supplies d.c. field excitation for the rotor. The load then can be coupled to the motor.

Single-Phase Motors. *Single-phase (1-θ) motors* are so called because their field windings are connected directly to a single-phase source. These motors are used extensively in fractional horsepower sizes in commercial and domestic applications. The advantages of using single-phase motors in small sizes are that they are less expensive to manufacture than other types, and they eliminate the need for 3-phase a.c. lines. Single phase motors are used in fans, refrigerators, portable drills, grinders, and so forth.

A single-phase induction motor with only one stator winding and a cage rotor is like a 3-phase induction motor with a cage rotor except that the single-phase motor has no magnetic revolving field at start and hence no starting torque. However, if the rotor is brought up to speed by external means, the induced currents in the rotor will cooperate with the stator currents to produce a revolving field, which causes the rotor to continue to run in the direction which it was started.

Several methods are used to provide the single-phase induction motor with starting torque. These methods identify the motor as split-phase, capacitor, shaded-pole, and repulsion-start induction motor.

Another class of single-phase motors is the a.c. series (universal) type. Only the more commonly used types of single-phase motors are described. These include the (1) split-phase motor, (2) capacitor motor, (3) shaded-pole motor, (4) repulsion-start motor, and (5) a.c. series motor.

Split-Phase Motor. The *split-phase motor* (see figure 9.100), has a stator composed of slotted lamination that contain a starting winding and a running winding.

✔ Note: If two stator windings of unequal impedance are spaced 90 electrical degrees apart but connected in parallel to a single-phase source, the field produced will appear to rotate. This is the principle of *phase splitting*.

The starting winding has fewer turns and smaller wire than the running winding, hence has higher resistance and less reactance. The main winding occupies the lower half of the slots and the starting winding occupies the upper half. When the same voltage is applied to both windings, the current in the main winding lags behind the current in the starting winding. The angle θ between the main and starting windings is enough phase difference to provide a weak rotating magnetic field to produce a starting torque. When the motor reaches a predetermined speed, usually 75% of synchronous speed, a centrifugal switch mounted on the motor shaft opens, thereby disconnecting the starting winding.

Because it has a low starting torque, the fractional-horsepower split-phase motor is used in a variety of equipment such as washers, oil burners, ventilating fans, and woodworking machines. The direction of rotation of the split-phase motor can be reversed by interchanging the starting winding leads.

Capacitor Motor. The *capacitor motor* is a modified form of split-phase motor, having a capacitor in series with the starting winding (see figure 9.101). The capacitor motor operates with an auxiliary winding and series capacitor permanently connected to the line. The capacitance in series may be of one value for starting and another

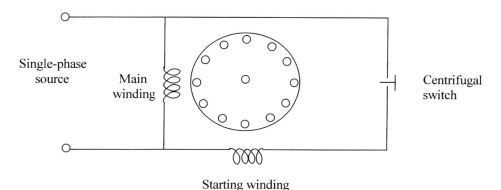

Figure 9.100.   Split-phase motor.

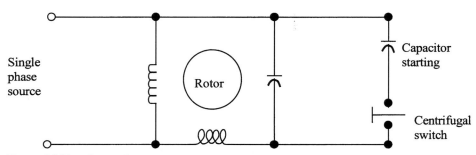

Figure 9.101.   Capacitor motor.

value for running. As the motor approaches synchronous speed, the centrifugal switch disconnects one section of the capacitor.

If the starting winding is cut out after the motor has increased in speed, the motor is called a *capacitor-start motor*. If the starting winding and capacitor are designed to be left in the circuit continuously, the motor is called *capacitor-run motor*. Capacitor motors are used to drive grinders, drill presses, refrigerator compressors, and other loads that require relatively high starting torque. The direction of rotation of the capacitor motor may be reversed by interchanging the starting winding leads.

Shaded-Pole Motor. A *shaded-pole* motor employs a salient-pole stator and a cage rotor. The projecting poles on the stator resemble those of d.c. machines except that the entire magnetic circuit is laminated and a portion of each pole is split to accommodate a short-circuited coil called a *shading coil* (see figure 9.102). The coil is usually a single band or strap of copper. The effect of the coil is to produce a small sweeping motion of the field flux from one side of the pole piece to the other as the field

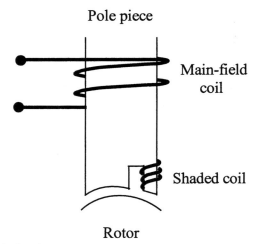

Figure 9.102.   Shaded pole.

pulsates. This slight shift in the magnetic field produces a small starting torque. Thus shaded-pole motors are self-starting. This motor is generally manufactured in very small sizes, up to 1/20 horsepower, for driving small fans, small appliances, and clocks.

In operation, during that part of the cycle when the main pole flux is increasing, the shading coil is cut by the flux, and the resulting induced emf and current in the shading coil tend to prevent the flux from rising readily through it. Thus, the greater portion of the flux rises in that portion of the pole that is not in the vicinity of the shading coil. When the flux reaches its maximum value, the rate of change of flux is zero, and the voltage and current in the shading coil are also zero. At this time the flux is distributed more uniformly over the entire pole face. Then as the main flux decreases toward zero, the induced voltage and current in the shading coil reverse their polarity, and the resulting mmf tends to prevent the flux from collapsing through the iron in the region of the shading coil. The result is that the main flux first rises in the unshaded portion of the pole and later in the shaded portion. This action is equivalent to a sweeping movement of the field across the pole face in the direction of the shaded pole. The rotor conductors are cut by this moving field and the force exerted on them causes the rotor to turn in the direction of the sweeping field. The shaded-pole method of starting is used in very small motors, up to about 1/25 hp, for driving small fans, small appliances, and clocks.

Repulsion-Start Motor. Like a d.c. motor, the *repulsion-start motor* has a form-wound rotor with commutator and brushes. The stator is laminated and contains a distributed single-phase winding. In its simplest form, the stator resembles that of the single-phase motor. In addition, the motor has a centrifugal device which removes the brushes from the commutator and places a short-circuiting ring around the commutator. This action occurs at about 75% of synchronous speed. Thereafter, the motor operates with the characteristics of the single-phase induction motor. This type of motor is made in sizes ranging from 1/2 to 15 hp and is used in applications requiring a high starting torque.

A.C. Series Motor

The *a.c. series motor* will operate on either a.c. or d.c. circuits. When an ordinary d.c. series motor is connected to an a.c. supply, the current drawn by the motor is low due to the high series-field impedance. The result is low running torque. To reduce the field reactance to a minimum, a.c. series motors are built with as few turns as possible. Armature reaction is overcome by using *compensating windings* (see figure 9.103) in the pole pieces.

As with d.c. series motors, in an a.c. series motor the speed increases to a high value with a decrease in load. The torque is high for high armature currents so that the motor has a good starting torque. A.C. series motors operate more efficiently at low frequencies.

Fractional horsepower a.c. series motors are called *universal motors*. They do not have compensating windings. They are used extensively to operate fans and portable tools, such as drills, grinders, and saws.

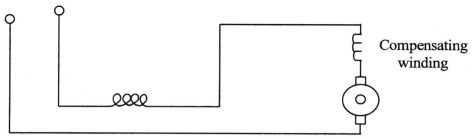

Figure 9.103.   A.C. series motor.

TRANSFORMERS

A *transformer* is an electric control device (with no moving parts) that raises or lowers voltage or current in an electric distribution system. The basic transformer consists of two coils electrically insulated from each other and wound upon a common core (see figure 9.104). Magnetic coupling is used to transfer electric energy from one coil to another. The coil that receives energy from an a.c. source is called the *primary*. The coil that delivers energy to an a.c. load is called the *secondary*. The core of transformers used at low frequencies is generally made of magnetic material, usually laminated sheet steel. Cores of transformers used at higher frequencies are made of powdered iron and ceramics, or nonmagnetic materials. Some coils are simply wound on nonmagnetic hollow forms such as cardboard or plastic so that the core material is actually air.

In operation, an alternating current will flow when an a.c. voltage is applied to the primary coil of a transformer. This current produces a field of force that changes as the current changes. The changing magnetic field is carried by the magnetic core to the secondary coil, where it cuts across the turns of that coil. In this way, an a.c. voltage in one coil is transferred to another coil, even though there is no electrical

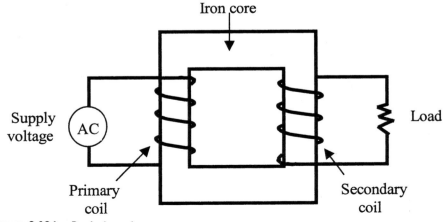

Figure 9.104.   Basic transformer.

connection between them. The number of lines of force available in the primary is determined by the primary voltage and the number of turns on the primary—each turn producing a given number of lines. If there are many turns on the secondary, each line of force will cut many turns of wire and induce a high voltage. If the secondary contains only a few turns, there will be few cuttings and low induced voltage. The secondary voltage, then, depends on the number of secondary turns as compared with the number of primary turns. If the secondary has twice as many turns as the primary, the secondary voltage will be twice as large as the primary voltage. If the secondary has half as many turns as the primary, the secondary voltage will be one-half as large as the primary voltage.

✔ **Important Point**: The voltage on the coils of a transformer is directly proportional to the number of turns on the coils.

A voltage ratio of 1:4 (read as "1 to 4") means that for each volt on the primary, there are 4 volts on the secondary. This is called a *step-up* transformer. A step-up transformer receives a low voltage on the primary and delivers a high voltage from the secondary. A voltage ratio of 4:1 (read as "4 to 1") means that for 4 volts on the primary, there is only 1 volt on the secondary. This is called a *step-down* transformer. A step-down transformer receives a high voltage on the primary and delivers a low voltage from the secondary.

POWER DISTRIBUTION SYSTEM PROTECTION

Interruptions are very rare in a power distribution system that has been properly designed. Still, protective devices are necessary because of the load diversity. Most installations are quite complex. In addition, externally caused variations might overload them or endanger personnel.

Figure 9.105 shows the general relationship between protective devices and different components of a complete system. Each part of the circuit has its own protective device or devices that protect not only the load, but also the wiring and control devices themselves. These disconnect and protective devices are described in the following sections.

Fuses

The passage of an electric current produces heat. The larger the current, the more heat is produced. In order to prevent large currents from accidentally flowing through expensive apparatus and burning it up, a *fuse* is placed directly into the circuit, as in figure 9.115, so as to form a part of the circuit through which all the current must flow.

✔ **Key Point**: A fuse is a thin strip of easily melted material. It protects a circuit from large currents by melting quickly, thereby breaking the circuit.

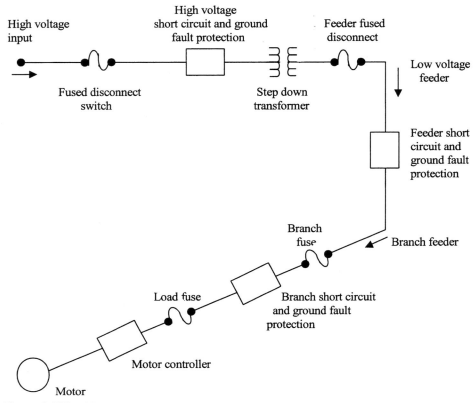

Figure 9.105.   Motor power distribution system.

The fuse will permit currents smaller than the fuse value to flow but will melt and therefore break the circuit if a larger, dangerous current ever appears. For instance, a dangerously large current will flow when a "short circuit" occurs. A short circuit is usually caused by an accidental connection between two points in a circuit that offer very little resistance to the flow of electrons. If the resistance is small, there will be nothing to stop the flow of the current, and the current will increase enormously. The resulting heat generated might cause a fire. However, if the circuit is protected by a fuse, the heat caused by the short-circuit current will melt the fuse wire, thus breaking the circuit and reducing the current to zero.

Fuses are rated by the number of amps of current that can flow through them before they melt and break the circuit. Thus we have 10-, 15-, 20-, 30-A fuses, and so on. We must be careful that any fuse inserted in a circuit be rated low enough to melt, or "blow," before the apparatus is damaged. For example, in a plant building wired to carry a current of 10 A it is best to use a fuse nor larger than 10 A so that a current larger than 10 A could never flow.

Some equipment, such as an electric motor, requires more current during starting than for normal running. Thus, fast-time or medium-time fuse rating that will give running protection might blow during the initial period when high starting current is required. *Delayed action* fuses are used to handle these situations.

Circuit Breakers

Circuit breakers are protective devices that open automatically at a preset ampere rating to interrupt an overload or short circuit. Unlike fuses, they do not require replacement when they are activated. They are simply reset to restore power after the overload has been cleared.

> ✔ **Key Point:** A circuit breaker is designed to break the circuit and stop the current flow when the current exceeds a predetermined value.

Circuit breakers are made in both plug-in and bolt-on designs. Plug-in breakers are used in load centers. Bolt-on's are used in panelboards and exclusively for high interrupting current applications.

Circuit breakers are rated according to current and voltage, as well as short circuit interrupting current. A single handle opens or closes contacts between two or more conductors. Breakers are single-pole, but can be ganged single-pole units—forming double-or-triple-pole devices opened with a single handle.

Several types of circuit breakers are commonly used. They may be thermal, magnetic, or a combination of the two. Thermal breakers are tripped when the temperature rises because of heat created by the overcurrent condition. Bimetallic strips provide the time delay for overload protection. Magnetic breakers operate on the principle that a sudden current rise creates enough magnetic field to turn an armature, tripping the breaker and opening the circuit. Magnetic breakers provide the instantaneous action needed for short circuit protection. Thermal-magnetic breakers combine features of both types of breakers.

Magnetic breakers are also used in circumstances where ambient temperature might adversely affect the action of a thermal breaker.

An important feature of the circuit breaker is its arc chutes, which enable the breaker to extinguish very hot arcs harmlessly. Some circuit breakers must be reset by hand, while others reset themselves automatically. When the circuit breaker is reset, if the overload condition still exists, the circuit breaker will trip again to prevent damage to the circuit.

Control Devices

Control devices are the electrical accessories (switches and relays) that govern the power delivered to any electrical load.

In its simplest form the control applies voltage to, or removes it from, a single load. In more complex control systems, the initial switch may set into action other control devices (relays) that govern motor speeds, servomechanisms, temperatures, and numerous other equipment. In fact, all electrical systems and equipment are controlled in some manner by one or more controls. A controller is a device or group of devices that serves to govern, in some predetermined manner, the device to which it is connected.

In large electrical systems, it is necessary to have a variety of controls for operation of the equipment. These controls range from simple pushbuttons to heavy-duty contactors that are designed to control the operation of large motors. The pushbutton is manually operated while a contactor is electrically operated.

Problems

9.1.   Is the charge of an electron positive or negative?
9.2.   Which have more free electrons: metal or insulators?
9.3.   Which is a semiconductor: copper, silicon, or neon?
9.4.   How many electron charges are there in the practical unit of one coulomb?
9.5.   How much potential difference is there between two identical charges?
9.6.   Define "difference in potential."
9.7.   The flow of 2 coulombs/second of electron charges is how many amps of current?
9.8.   Which has more resistance, carbon or copper?
9.9.   The same voltage applied, which resistance will allow more current, 4.9 Ω or 6000 Ω?
9.10. Calculate R for 12 volts and 0.006 A.
9.11. An electron heater takes 15 A from the 120-volt power line. Calculate the power.
9.12. How much is the R of a 100-watt 120-volt light bulb?
9.13. A series circuit has IR drops of 10 volts, 20 volts, and 30 volts. How much is the applied voltage E_T of the source?
9.14. Each of three equal resistances dissipate 2 watts. How much P_T is supplied by the source?
9.15. Parallel branch currents are 1 A for I_1, 2 A for I_2, and 3 A for I_3. Calculate I_T.
9.16. Find R_T for three 4.7-MΩ resistances in parallel.
9.17. How much is the IR voltage drop across a closed switch?
9.18. How much is the resistance of a good fuse?
9.19. Which type can be recharged, a primary or a secondary cell?
9.20. A.C. voltage varies in _____ and reverses in _____.
9.21. A sine-wave voltage has a peak value of 170 volts. What is its value at 45°?
9.22. State the law of attraction and repulsion of charged bodies.
9.23. What is true power?
9.24. In a series circuit whose source voltage is 30 volts, there are three resistors. The voltages across two of the resistors are 6 volts and 10 volts. What is the voltage across the third resistor?
9.25. Explain the difference between d.c. and a.c. current.
9.26. What is the L/R time constant of a circuit whose total inductance is 100 mh and whose total resistance is 50 Ω?

9.27. How much is Z_T for a 20-Ω R in series with a 20-Ω X_C?

9.28. How much is the RC time constant for 470 pF in series with 2 MΩ on charge?

9.29. What is the unit for apparent power?

9.30. In a capacitor, is the electric charge stored in the dielectric or in the metal plates?

Answers to Chapter Problems

CHAPTER 1: MEASUREMENT

1.1. A height of 6 ft 4 in.: $(6 \times 12) + 4 = 76$ in. $\times$ 2.54 cm/in. $= 193$ cm or 1.9 m

1.2. SA $= 4\pi^2 = 4$ (3.14) $(1,738 \text{ km})^2 = (3.79 \times 10^8 \text{ km}^2)$; 2/3 $(3.79 \times 10^8 \text{ km}^2) = 2.52 \times 10^8 \text{ km}^2$

1.3. 5 mL = 1; 5.2 g = 2; 5.0 kg = 2; 5,000 L = 4; 0.005 m = 1
    1×10^0; (b) 3×10^1; (c) 5.72×10^9; (d) -6.1×10^{-9}

1.4. 0.1

1.5. 1×106

1.6. Density $= 45 \text{ g}/15 \text{ cm}^3 = 3.0 \text{ g/cm}^3$

1.7. Density $= 60 \text{ g}/30 \text{ cm}^3 = 2.0 \text{ g/cm}^3$

1.8. SG $= 28 \text{ g/cm}^3 \div 10 \text{ g/cm}^3 = 2.8$

1.9. SG $= 174.8 \text{ g/cm}^3 \div 62.4 \text{ g/cm}^3 = 2.8$

CHAPTER 2: FORCE AND MOTION

2.1. Speed $=$ distance/time $= 375 \text{ m}/7 \text{ hr} = {\sim}54$ mph

2.2. Time $=$ distance/velocity $= (1 \text{ min})/(15 \text{ mi/h}) = 0.06$ h, or ${\sim}4$ min

2.3. Pythagorean theorem: speed $= \sqrt{([520 \text{ mph}]^2 + [50 \text{ mph}]^2)} = {\sim}522$ mph, to the southeast

2.4. $0 = 0 \text{ m/x} + (9.8 \text{ m/s}^2)(12 \text{ m})(60 \text{ s/min}) = 7.1 \times 10^3 \text{ m/s}$, or 7.1 km/s

2.5. $= 60 \text{ m} + (0 \text{ m/s})t + (-9.8 \text{ m/s})t^2$, $t = \sqrt{60 \text{ m} /9.8 \text{ m/s}^2} = 2.5$ s

2.6. Magnitude, direction

2.7. Third law of motion

CHAPTER 3: WORK, ENERGY, AND MOMENTUM

3.1. Velocity (because it is squared in the formula)

3.2. Work $= (250 \text{ N}) (15 \text{ m}) = 3750 \text{ N·m} = 3750 \text{ M} = 3.75$ kj

3.3. PE $= (1000 \text{ kg}) (9.8 \text{ m/s}^2) 10 = 9.8 \times 10^4$ J; $(4 \text{ kg}) (9.8 \text{ m/s}^2) (200) = 7.8 \times 10^4$ J; the work required is equal to the PE of each object.

3.4. Assuming all his potential energy is converted to kinetic energy, PE + KE (top) $=$ PE + KE (bottom); $(2.2 \text{ m})(9.8 \text{ m/s}^2)(30 \text{ kg}) + 0 = 0 + \frac{1}{2} (30 \text{ kg}) v^2$; $v = \sqrt{43.1 \text{ m}^2/\text{s}^2} = 6.5$ m/s

3.5.   Work = force × distance = (12,000 kg) (9.8 m/s²) (0.6 m) = 7.1 × 10⁴ J; power = work/time = (7.1 × 10⁴)/(30 min) (60 s/min) = 39.4 J/s = 39.4 W

CHAPTER 4: CIRCULAR MOTIONS AND GRAVITY

4.1.   Twice as much
4.2.   180°; π rad; 200 × 360° = 72,000; 200 × 2π = 400π rad
4.3.   (45 ev/min) (2π rad/rev) (1 min/60 s) = 4.7 rad/s
4.4.   2 πrad/5 s = 0 + angular acceleration (5 s); ang. accel. = 0.25 rad/s²
4.5.   Less

CHAPTER 5: ATOMS, MOLECULES, AND ELEMENTS

5.1.   Protons, +1; electrons, −1; neutrons, 0
5.2.   An atom comprises a nucleus (containing neutrons and protons) and electrons. A molecule is a collection of atoms, held together by chemical bonds.
5.3.   Protons
5.4.   Neutron
5.5.   Electron

CHAPTER 6: THERMAL PROPERTIES AND STATES OF MATTER

6.1.   4 Celsius degrees (80/20)
6.2.   100 Btu
6.3.   14,000 Btu of energy can be exchanged each hour by the device.
6.4.   110 cal
6.5.   They lack water vapor to maintain temperatures.

CHAPTER 7: WAVE MOTION AND SOUND

7.1.   Wavelength = 3 m; period = 1 s; velocity 3 m/s
7.2.   Yes; at higher air temperatures molecular collisions are more frequent, and sound is transferred through these sorts of collisions (longitudinal waves).
7.3.   Longitudinal
7.4.   Amplitude
7.5.   100,000 times

CHAPTER 8: LIGHT AND COLOR

8.1. Time = (5 mi)/(183,310 mi/h)(3600 s/h) = 0.098 s
8.2. 20 ft-c = C/(40 ft)²; C = 32,000 candles
8.3. Frequency = c/wavelength = (3 × 10⁸ m/s)/(s/656 × 10²⁹ m) = 4.6 × 10¹⁴ Hz
8.4. Frequency = c/wavelength = (3 × 10⁸ m/s)/(0.20 m) = 1.4 × 10⁹, or 1.4 GHz
8.5. Intensity of light

CHAPTER 9: ELECTRICITY

9.1. Negative
9.2. Metals
9.3. Silicon
9.4. 6.25×10^{18}
9.5. Zero
9.6. The force that causes free electrons to move in a conductor as electric current.
9.7. 2 A
9.8. Carbon
9.9. 4.9 Ω
9.10. 3 A
9.11. 1.8 kW
9.12. 144 Ω
9.13. 60 volts
9.14. 6 watts
9.15. $I_T = 6$ A
9.16. $R_T = 1.57$ MΩ
9.17. Zero
9.18. Zero
9.19. Secondary
9.20. Magnitude; polarity
9.21. 120 volts
9.22. Like charges repel each other and unlike charges attract each other.
9.23. Power that is dissipated as heat.
9.24. 14 volts
9.25. D.C. flows in only one direction, while a.c. periodically reverses the direction of flow.
9.26. 2 milliseconds (divide 100 mh by 50 Ω)
9.27. 28.28 Ω
9.28. 940 μs
9.29. Volt-amp
9.30. Dielectric

Recommended Reading

Cutnell, J. D., and K. W. Johnson. *Physics*. 7th ed. New York: Wiley, 2006.

Giancoli, D. D. *Physics: Principles and Applications*. 5th ed. New York: Reason Education, 2004.

Gibson, J. *The Ultimate Physics Tutor*. DVD-video, full screen. New York: Tapeworm, 2006.

Halliday, D., R. Resnick, and J. Walker. *Fundamentals of Physics*. 7th ed. New York: Wiley, 2004.

———. *Fundamentals of Physics Extended*. New York: Wiley, 2007.

Hewitt, P. G. *Conceptual Physics*. 10th ed. Boston: Addison Wesley, 2005.

Holzner, S. *Physics for Dummies*. New York: Wiley, 2004.

———. *Physics Workbooks for Dummies*. New York: Wiley, 2007.

Kuhn, K. F. *Basic Physics: A Self-Teaching Guide*. New York: Wiley, 1996.

Serway, R. A., J. S. Faughn, and C. Vuille. *College Physics*. 8th ed. New York: Brooks Cole, 2008.

Walker, T. E. *Physics*. 3rd ed. Boston: Benjamin Cummings, 2006.

Young, H. D., and R. A. Freedman. *University Physics with Modern Physics with Mastering Physics*. Boston: Addison Wesley, 2007.

Index

About the Author

Frank R. Spellman is assistant professor of environmental health at Old Dominion University in Norfolk, Virginia. He holds a BA in public administration, a BS in business management, an MBA, and an MS and a PhD in environmental engineering.

Spellman consults on homeland security vulnerability assessments for critical infrastructure, including water/wastewater facilities nationwide. He also lectures on homeland security and health and safety topics throughout the country and teaches water/wastewater operator short courses at Virginia Tech.

Spellman's work has been cited in more than four hundred publications and he is a contributing author for *The Engineering Handbook*, second edition. He has written more than fifty books that cover topics in all areas of environmental science and occupational health, including *Chemistry for Nonchemists* (2006), *Biology for Nonbiologists* (2007), and *Ecology for Non-Ecologists* (2008), all published by Government Institutes.